NEW LABOUR AND PLANNING

Following the Thatcher and Major administrations there was an apparent renaissance of planning under New Labour. After a slow start in which Labour's view of planning owed more to a neoliberal, rolled back state model reminiscent of the New Right, the government began to appreciate that many of its wider objectives, including economic development, climate change, democratic renewal, social justice and housing affordability, intersected with and were critically dependent upon the planning system. In England a new system of development plans was created, along with the notion of 'spatial planning', as a way of bringing together the fragmented landscape of governance towards a range of broad objectives, including sustainable development, urban renaissance and tackling climate change.

A wide range of initiatives, management processes, governance vehicles and policy documents emanated from government. Planning, like other areas of the public sector, was to be reformed and modernised and given a prime role in tackling national, high profile priorities such as increasing housing supply and improving economic competitiveness. The result was a hyperactive period of activity and change that had a variety of intended and unintended impacts as well as longer term implications for the way in which we think about planning and the role of the state.

But the experiences of Labour tell us more than how national governments succeed, or don't, in policy change. The Labour era also calls to attention the nature of planning itself and how sources of stability and change in the wider governance landscape react to and interpret change. Drawing upon an institutionalist framework, the book also seeks to understand how and in what circumstances change emerges, either in an evolutionary or a punctuated way. It will, for the first time, chart and explore the changing nature of development and planning over the Labour era while also stepping back and reflecting upon what such changes mean for planning generally and the likely future trajectories of reform and spatial governance.

Phil Allmendinger is a Fellow of Clare College and Professor of Land Economy at Cambridge University, UK.

NEW LABOUR AND PLANNING

From New Right to New Left

Phil Allmendinger

Routledge
Taylor & Francis Group
LONDON AND NEW YORK

First published 2011
by Routledge
2 Park Square, Milton Park, Abingdon, Oxon, OX14 4RN

Simultaneously published in the USA and Canada
by Routledge
270 Madison Avenue, New York, NY 10016

Routledge is an imprint of the Taylor & Francis Group, an informa business

© 2011 Phil Allmendinger

The right of Phil Allmendinger to be identified as author of this work has been asserted by him in accordance with sections 77 and 78 of the Copyright, Designs and Patents Act 1988.

Typeset in Bembo by
Book Now Ltd, London
Printed and bound in Great Britain by
TJ International Ltd, Padstow, Cornwall

All rights reserved. No part of this book may be reprinted or reproduced or utilised in any form or by any electronic, mechanical, or other means, now known or hereafter invented, including photocopying and recording, or in any information storage or retrieval system, without permission in writing from the publishers.

British Library Cataloguing in Publication Data
A catalogue record for this book is available from the British Library

Library of Congress Cataloging-in-Publication Data
Allmendinger, Philip, 1968–
 New Labour and planning : from New Right to New Left / Philip Allmendinger.
  p. cm.
 Includes bibliographical references and index.
 1. Regional planning—Great Britain. 2. Labour Party (Great Britain) I. Title.
 HT395.G7A55 2011
 324.24107—dc22   2010031309

ISBN13: 978–0–415–59748–7 (hbk)
ISBN13: 978–0–415–59749–4 (pbk)
ISBN13: 978–0–203–83199–1 (ebk)

For Claudia, Hannah, Lucia and Eleanor

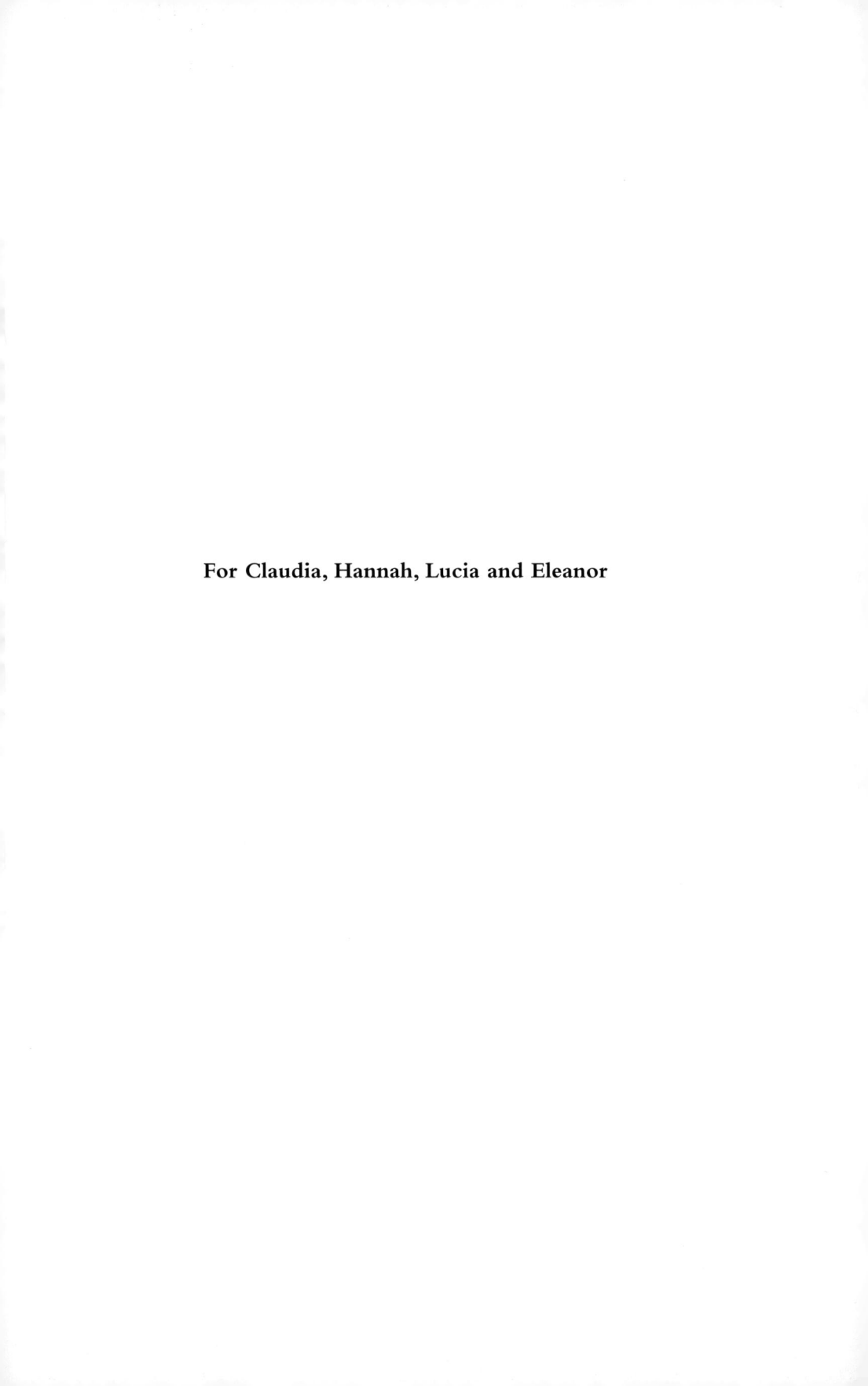

CONTENTS

List of illustrations		*ix*
Preface		*xi*
Acknowledgements		*xiii*
1	From New Right to New Left	1
2	New Labour and planning	15
3	Understanding planning under Labour	38
4	Planning and urban policy	62
5	Spatial planning	88
6	Hitting the target and missing the point: speed in planning decisions	112
7	Development, infrastructure and land taxation	130
8	Conclusions	152
Notes		*169*
References		*171*
Index		*187*

ILLUSTRATIONS

Figures

5.1	Organisational arrangements for growth management in Cambridge	102
6.1	Application- and site-based performance measures of the planning process	120

Tables

1.1	Characterising the New Right's approach to planning	9
3.1	Areas where local planning authorities have been given additional or more complex responsibilities since the Town and Country Planning Act 1990	51
3.2	A typology of planning styles	56
3.3	Characteristics of six planning styles	57
4.1	Selected key events in King's Cross redevelopment	73
4.2	Selected key events in Radstock regeneration scheme	81
4.3	Selected policy context for Radstock scheme	82
6.1	Development control time for individual sites	121
6.2	Planning process time by local authority and development control performance, 2006	122
8.1	A New Institutionalist understanding for change in planning	163

Maps

4.1	The King's Cross Redevelopment Area	72
4.2	The Norton Radstock Regeneration Area	77
5.1	Cambridge sub-region development proposals and administrative boundaries	99
5.2	The Thames Gateway area	105

PREFACE

This book is an account of the changes to planning during the New Labour era. Despite drawing upon a range of case studies, it will never be possible to capture fully the experiences and practices of planning over a relatively long period and across a diversity of places. The book is therefore and inevitably a partial and personal account. Those wishing to read a vituperative take on the New Labour years will be disappointed. Similarly, those who felt that planning emerged strengthened and reborn will also be frustrated. While I have attempted a balanced understanding, the theme running through the book is itself one of disappointment: a disappointment with the New Labour government and a disappointment with planning and planners. I include planning academics and the professional institute in that last category. Despite the investment of time and resources, and against a backdrop of significant development demand, the opportunity to reinvent planning was lost. The proposals emerging from the coalition government on planning can be summarised as 'less is more'. They may be ideologically driven but they can be justified and defended by reference to experiences of the previous thirteen years.

New Labour was guilty of seeking to achieve too much and of not understanding how the system worked. But it was also highly dependent upon planners themselves to effect change. There are plenty of examples of innovative practices and outcomes. Equally, old practices die hard. Leadership and vision were lacking. The new totem of spatial planning advanced by the profession and academics was a chimera that, to paraphrase Aaron Wildavsky, was everything and nothing. It echoed New Labour's own obsession with positive messages and inclusive discourses around sustainable development. Conflict and difficult choices, once the role of planning, were either replaced by consensus, partnership-based governance or passed on to multiple, arm's-length enquiries. The depoliticisation of planning was merely a postponement of conflict. There cannot be 'win–win–win' solutions in the messy, unequal and divergent real world. And so it turned out to be. By the time that this was realised it was too late. From 2008 the implications of the credit crunch on development activity began to be felt. Attention shifted, and planning was reorientated towards economic development and competitiveness.

The end of New Labour came in May 2010. There are a number of possible futures for planning under the coalition government. What planners and others need is to stand back and reflect critically upon the experiences and lessons of the New Labour era so as to help shape the future.

ACKNOWLEDGEMENTS

This book emerged over a long period and draws upon a wide range of research, some of which was undertaken with others. Particular acknowledgement and thanks should go to Graham Haughton and Mark Tewdwr-Jones, who have helped shape what follows in a number of ways. Jeremy Smalley has provided an invaluable sounding board and insightful input on my take on the evolution of planning over the years as well as being a great friend. Alan Prior and Cliff Hague at Heriot-Watt University helped shape my thoughts on planning when I was a student and then as a colleague.

Numerous others provided their time and input generously, and I would like to express my sincere thanks to: Simon Payne, John Summers and David Roberts at Cambridge City Council; Alex Plant and John Onslow of Cambridgeshire Horizons; Peter Studdert at South Cambridgeshire District Council; Robert Evans of Argent; Paul Wilmott, C. B. Richard Ellis and Adrian Gurney from Arup; Diana Chin, Pete Tyler, Helen Hartley, Kelvin MacDonald and Colin Lizieri at the University of Cambridge; Derek Hooper and Kate Mack of the Norton Radstock Regeneration Company; James Buxton of Bidwells; Rebecca McAllister at Hives Planning; Pat McAllister, Franz Fürst, Michael Ball and Kathy Hughes at the University of Reading; and Mark Oranje at the University of Pretoria.

1
FROM NEW RIGHT TO NEW LEFT

Introduction

Views on New Labour, as with the preceding Conservative administrations, are seldom ambiguous. It is difficult to be detached and assess New Labour's impact and legacy objectively when issues such as the Iraq war continue to cast a long shadow. Putting this to one side, there are other problems even if we focus upon planning, and there are a number of possible narratives concerning the experiences of planning under Labour. One view might concern a 'renaissance of planning' and how, under Labour, it was again in the ascendency. New challenges and issues around climate change and sustainable development coupled with the need for multi-scalar and cross-sectoral working provided a heightened importance for planning through the notion of 'spatial planning'. By 2010 planning had emerged from the New Labour era in a position of strength and influence, charged with delivering and coordinating important elements of the government's objectives. Another possible view could be of 'business as usual': planning remained unreconstructed, slow and cumbersome. Despite the opportunity and substantial additional resources, planning failed to evolve from a 'command and control' form of regulation and embrace more networked and flexible forms of governance. A further perspective could highlight the democratic deficit within planning as it became a form of neoliberal, spatial governance, paying lip service to the wishes of local communities and attempting to force through change and development. Planners became complicit in the 'roll out' of neoliberalism, working closely with new, unelected and unaccountable bodies such as Regional Development Agencies and the Infrastructure Planning Commission to bypass local authorities and people. Yet another view could be that planning survived an inchoate scattergun of policies and initiatives based upon competing and irreconcilable objectives for and views of planning. Emphases upon increased public involvement and speed of decision-making, a 'step-change' in housing delivery and the protection of green belts, sustainable development and economic competitiveness provided an unachievable framework for planning and planners. These and other objectives, combined with systemic reform – particularly in the 2004 Planning and Compulsory Purchase Act – and a development boom, meant that the inevitable outcome was always likely to be failure.

It would be wrong to portray these possible views as mutually exclusive. There is a cognitive dissonance around the nature of planning during the Labour era: all of these seemingly contradictory narratives can be justified through recourse to evidence and all could be held simultaneously. Any assessment of the changes to planning over thirteen years depends upon normative positions around the role of the state in general and planning in particular, the issue in question and the point in time. Labour became serious about planning and climate change only well into its second term and was initially concerned with scaling it back. Assessments also depend to a large degree on what we mean by 'planning'. Planning has always been a diffuse phenomenon but was even more so under New Labour, as planners became a kind of coordinating, networking 'bridge' between different policy sectors and the public, private and voluntary sectors, all operating at multiple scales in the fractured, networked landscape of governance. 'Planning' also differs if the focus is upon nationally significant infrastructure, on the one hand, or a house extension, on the other.

In addition to the normative underpinnings of any assessment, the issue in question, the period and what is taken to include planning, an equally important and often overlooked influence upon change is the inherited legislative and policy framework and the policy trajectory of the previous administration. Those who concentrate solely on planning could be forgiven for overestimating its significance in the thinking of Labour when in opposition and in government. The truth is that, during the 1980s and 1990s, Labour showed little interest in planning other than as a largely middle-class mechanism for thwarting housing development and economic growth (McCormick, 1991). The 1997 Labour manifesto talked of putting the environment at the heart of policy-making and developing an integrated transport system. By contrast, by 2010 the Labour and Conservative manifestos contained a great deal more on planning, particularly in relation to housing delivery, environmental protection and climate change. The 1998 Planning White Paper (DETR, 1998) contained little, if any, vision for planning, while the 2001 Green Paper on planning (DTLR, 2001a), issued four years after Labour came to power, was more forthright on what role the government perceived for planning, though this was a very much scaled-down, market-supportive function. I go into more detail on environmental thinking and policy within Labour in the next chapter. Suffice to say here that Labour took a number of years before it began to become interested in planning as anything more than a brake upon competitiveness and growth.

Against a backdrop of such difficulties and issues, the question that arises is: Why bother? The issues above are not unique to New Labour and characterise the complexity and context of contemporary policy studies. The main difference between assessments of the New Right and the New Left largely concern the starting point. Evaluations of Thatcherism usually began by identifying an ideology. As I go on to discuss in Chapter 2, there is much less agreement on the existence of 'Blairism', never mind what it entailed. If we want to understand change, then complexity and contingency provide the backdrop to any study. The second main reason is that, despite such contingency, we can still learn. New Labour was renowned for being concerned less with ideology and more with 'what works'. Its approach to public policy was distinct in many ways, particularly the focus upon cross-cutting issues, such as health, rather than policy sectors, such as health care. In such an approach the means are less important. Planning was a 'tool' to achieve outcomes such as social inclusion, addressing climate change and improving economic competitiveness. If planning 'didn't work', then there was nothing sacred about it. It could be replaced by other 'tools' or approaches. A variety of policy initiatives were introduced in order to achieve such wider objectives, among them the post-2004 system of development planning, the Planning Gain Supplement/Community Infrastructure Levy and the notion of spatial planning, among others. The ways in which these initiatives played out, evolved and effected change tell us as much about contemporary governance as they do about planning itself.

Such pragmatism is not the starting point, however. The starting point is that New Labour did not arrive in May 1997 ready formed to take forward a planning agenda. In fact, it was woefully unready, its priorities being in other policy areas such as education and health. Labour's approach emerged and then evolved, taking in a broad range of ideas and influences. Among the most significant of these during its first term was the inheritance and trajectory from the previous administration, which, according to some, it resembled in many ways. This chapter provides an overview and analysis of that inheritance as a way of helping better to understand the nature of New Labour and its approach to planning in Chapter 2.

From New Right to New Left

The stance of the Labour Party in opposition and government was not the only influence upon change. Labour did not inherit a *tabula rasa* but a system that had been subject to attempts to impose change and an extant policy and institutional framework. The election of the first Thatcher government in 1979 represented a clear break with the postwar social democratic consensus and the initiation of a New Right (Cloke, 1992). There were distinct approaches to planning within the broad umbrella of the New Right, the two most significant eras being between the periods 1979 to 1990 and 1990 to 1997.

The New Right: 1979–1990

Part of the governing ideology of the Thatcher era was the notion that 'There Is No Alternative': Britain's ills needed to be addressed head-on by a radical leader and government committed to market-based solutions, the authority of the state and personal freedom. In time this governing ideology has slid into mythology, as Thatcher promulgated the notion that she would not make a U-turn from the unpopular but correct course her government was pursuing. The coherence of the Conservative programme in the 1980s was less lucid than presented, both ideologically and in practice, and the tension arose from the fusion of two distinct collections of ideas that underpinned Conservative thinking during the 1970s and 1980s. The New Right was founded upon two sets of ideas that revolved around how the economy should be organised and the style and content of government (Thornley, 1991). These two strands have been variously labelled social market economy and authoritarian popularism (Gamble, 1984), free economy and strong state (Gamble, 1988), economic liberalism and authoritarianism (Edgar, 1983), neoliberalism and combative Toryism (Norton and Aughey, 1981) and liberalism and Conservatism (King, 1987). However they are labelled, all analyses point to the attempt to move Britain towards a freer, more competitive, more open economy and a more repressive, more authoritarian (and centralised) state (Gamble, 1984: 8). The fusion of these two central tenets was largely a paper exercise, as the experience of planning demonstrated.

The authoritarian tenet of Thatcherism privileged centralisation and the neoliberal tenet required a deregulated market. Both could be reconciled through the broad strategy of 'rolling back the state' and minimising local discretion. However, there were numerous areas of public policy where there was no obvious approach that would satisfy both tenets. In planning one could point to environmental and conservation concerns as prime examples of different solutions to the same problem from within government. The neoliberal approach pointed towards deregulation and market orientation of policy in conservation, while, to authoritarians, issues such as identity were closely bound up with physical characteristics of settlements – the mythical 'green and pleasant land', for example. For authoritarians, preserving such reflections and influences on the national psyche was important. One of the consequences for a radical government intent on some change was the need to phrase policy and legislation in such a way as to satisfy both camps. The result was vague and ambiguous policy objectives and guidance. Such an approach to planning was ideologically patchy and often framed in an indistinct way so as to placate different streams within the party.

Policy ambiguity did not, however, detract from the rhetorical attacks upon planning, which were almost universally hostile. As secretary of state for planning Michael Heseltine put it, thousands of jobs are locked away every night in the filing trays of planning departments. High profile initiatives such as Urban Development Corporations (UDCs) and Enterprise Zones (EZs) played up to this anti-planning, deregulatory rhetoric, though their actual impact, such as the massive physical changes in the London docklands, was related far more to the significant fiscal and financial incentives involved (Allmendinger and Thomas, 1998). The impact of initiatives such as that in the docklands, while disputed (Brownill, 1990) and in large part dependent upon the serendipity of financial deregulation in the City of London and the consequent demand for commercial floorspace, went some way towards 'winning the argument' that planning controls were a 'burden on business' (HM Government, 1985). Using the 'success' of the deregulatory elements of planning in EZs and UDCs, the second Thatcher government introduced a range of initiatives that sought to roll back the scope of planning controls, among them Simplified Planning Zones (SPZs) (Allmendinger, 1998), the B1 Use Class and a Green Paper entitled *The Future of Development Plans* (DoE, 1986). The Green Paper proposed a unitary system of development plans to replace the two-tier approach of structure and local plans. Criticism of development plans focused upon the lengthy time to prepare them, with only a small minority of authorities having adopted a plan by 1988 (Thornley, 1991). Critiques of the existing system were not confined to delays but also sought to engage with more fundamental issues around their purpose:

> Structure plans are often too long and contain irrelevant and over-detailed policies; the relationship between structure and local plans is unsatisfactory partly because of the way in which their procedures are interlocked and partly because their contents overlap; and the procedures for preparing structure and local plans are too complex.
>
> *(McConnell, 1987: 95)*

The proposals for reform met with a broadly positive response (though not from the county councils), mainly as, when the anti-planning rhetoric was removed, they were seen as pragmatic rather than dogmatic.

While there was a clear mismatch between deregulatory rhetoric and proposals, the more conservative, authoritarian-inspired changes actually had a greater impact. Some of the high profile initiatives for planning, such as UDCs and EZs, proclaimed their deregulatory credentials, though they were concerned as much with centralisation and a diminution of local discretion. Centralisation included a bolstering of the role of the secretary of state through increasing the scope and significance of central policy while reducing local discretion to ignore it. For example, changes to central government planning guidance sought to reduce the scope of discretion and reorientate planning considerations so as to reduce subjective judgements concerning design and appearance (Allmendinger and Thomas, 1998). Centralisation and the reduction of discretion formed an important theme in both New Right thinking and the approach to planning and in some ways complemented the liberalisation and deregulation tenets. Comparisons with US and continental European zoning-based approaches to planning, for example, highlighted how continental European legally binding as opposed to the UK's indicative plans increased certainty for developers and communities and resulted in quicker decisions (Thornley, 1991). EZs and SPZs sought to emulate such zoning-based systems and involved, on paper at least, a reduction in local planning controls, an increased role for the secretary of state and more certainty for the market (Allmendinger, 1998).

The attempt to 'roll up' planning controls and introduce a more market-based approach came up against two main problems. First, political and electoral impacts provided a much stronger steer on policy than either liberal or authoritarian ideology. Reaction to changes to planning did not play well in many, largely Conservative voting areas, where there was a resistance to new development and scepticism towards new approaches. The emergence of a number of proposals for new settlements in the south-east of England by a consortium of housebuilders was met with fierce resistance by local residents and led some Conservative MPs to see the need for a strong planning system to focus necessary new development elsewhere (Ward, 2004). Around the same time there was a growing awareness of green issues and sustainable development, particularly following the publication of the Bruntland Commission report (Bruntland, 1987) and the European Commission's Green Paper on the urban environment (CEC, 1990). In 1989 the Green Party secured nearly 15 per cent of the vote in the European elections, bringing environmental issues, including planning controls, to popular and political attention. The perception of planning shifted from being a bureaucratic impediment to growth and competitiveness to an essential plank of environmental policy.

Second, the property industry was not embracing deregulation with anything like the enthusiasm that the New Right had envisaged. SPZs allowed landowners or developers to request that a local planning authority replace the discretionary, 'plan and permission' approach in their areas with a zoning-based system. If a local planning authority refused, the secretary of state could impose one. However, only a handful of privately initiated SPZs were proposed and even fewer were eventually introduced. The main reasons were that landowners and developers perceived advantages in the discretionary approach through being able to know what would be proposed and being able to comment or object. A zoning-based system removed that option and, while providing certainty that a specified range of uses would be permitted, also created uncertainty over which eventual use and building form would emerge. This amounted, in the conclusion of Allmendinger and Thomas (1998: 240), to a 'spectacular misreading of the market supportive role of planning'. Property interests were simply unwilling to see a wholesale deregulation of planning. The viability of investments depended to a high degree on the certainty provided by planning regulation.

Against this evolving backdrop, the 1986 Green Paper emerged in 1989 as a White Paper, also entitled *The Future of Development Plans* (DoE, 1989). The proposals remained similar, though the ends to which the new system would be put had shifted. The publication a year later of the White Paper *This Common Inheritance* (DoE, 1990) pointed towards a purpose for planning that went beyond a market-supportive role to one around environmental stewardship. As part of this the proposal to abolish structure plans was dropped and, instead, the latter were to provide a strategic vision for the now mandatory local plans. As a consequence, the 1990 Town and Country Planning Act consolidated a range of changes introduced since the last major planning Act in 1971. Thus, over a decade after the

Conservatives came to power with a radical agenda, planning found itself far from minimised and serendipitously allied with growing concerns around the environment and climate change.

The New Right: 1990–1997

The replacement of Margaret Thatcher as prime minister by John Major in 1990 underscored the shift in attitude towards the role of the state and planning that had emerged in the previous decade (Allmendinger and Tewdwr-Jones, 1997). Not only was the attitude of Major important in shaping a less antagonistic approach but there were other factors that helped influence the situation. The Conservative Party's reduced parliamentary majority following the 1992 general election combined with a very public schism over Europe meant that Major was less able to force through unpopular changes and was more reliant upon conciliation than confrontation (Marsh and Rhodes, 1992). The emergence of the environment and climate change as national issues was accompanied by a shift from a focus on economic policy in the 1980s to a concern with more complex social problems, including education, crime and health (Kavanagh, 1994). Following the decision to abolish the poll tax, local government was also of prime importance, with a major review of local government boundaries and functions announced as part of an agenda to improve the responsiveness and accountability of public services (HM Government, 1991).

This period witnessed the consolidation of the shift in approach and attitude towards planning that had emerged towards the end of the 1980s. It is tempting to view the changes individually, though this would be to miss the overall impact, as the totality of the separate alterations was greater than the sum of the parts. First, a key change came in the form of the 1991 Planning and Compensation Act and its insertion of an amendment into the 1990 Act. The new Section 54A of the 1990 Act stated that 'Where, in making any determination under the planning Acts, regard is to be had to the development plan, the determination shall be made in accordance with the plan unless material considerations indicate otherwise' (Planning and Compensation Act, 1991: Section 26).

While the impact of the new wording has been questioned (e.g. MacGregor and Ross, 1995; Gatenby and Williams, 1996), the main and intended outcome was that planners, the public and the development industry should place more emphasis upon the plan as the strategy for new development in an area. This was a distinct departure, if not a U-turn, from the point of view taken in the 1986 Green Paper and the general attitude during the 1980s. The idea that if a proposal did not conform to the plan then it should not proceed represented a swing towards a more zoning-based approach (Allmendinger, 2006). However, the reinforcement of the plan-led approach was far from a form of devolution or localism. The development plan had conditional primacy under Section 54A 'unless material considerations indicate otherwise'. A significant and determinate material consideration was whether the plan accorded with national planning policy. Two issues arose from this. First, while national planning policy does not have the same weight for decision-makers as delegated legislation or statutes, it does constitute guidance in plan preparation and decision-taking that should be taken into account. Another consideration was whether the development plan was up to date, and this included the extent to which it reflected national policy. Between 1988 and 1996 the government issued thirty-two Planning Policy Guidance Notes (PPGs), ten of which were revised PPGs (Allmendinger and Tewdwr-Jones, 1997). Second, because of the increased significance following the 1991 Act, the time taken to prepare plans increased as objectors, developers and others focused their attention upon them (Edwards and MacCafferty, 1992; Taussik, 1992; Lavers and Webster, 1994). The outcome was that the development plan was rarely up to date or even in place when proposals were assessed, further strengthening the importance of national planning policy. On the face of it, Section 54A strengthened the significance of the development plan but actually constituted a centralisation of control and policy, representing a shift towards the more authoritarian strand in New Right thinking.

A second key theme of planning under the Major government was the focus upon performance in development control. The publication of the Citizens' Charter in 1991 heralded the beginning of a fixation with performance and targets in the public sector as part of the focus upon improving services and introducing market mechanisms into public-sector management. The concern with speed of decision-making in development control went back to the 1970s, though the Major era witnessed a step-change in the scope and use of performance indicators. A related dimension was the introduction of a form of privatisation of some local authority services through Compulsory Competitive Tendering, which aimed to introduce competition to the provision of public services. Underperforming sectors within local government could be 'contracted out' (privatised), thereby providing an incentive to meet targets and improve performance. The concentration on performance constituted another arm of increased centralisation and the minimisation of local discretion, complementing the impacts of Section 54A.

The third theme concerned the introduction of competition for resources and the increased emphasis upon partnerships, particularly through the City Challenge programme, which ran from May 1991, and the Single Regeneration Budget (SRB) from 1994. Along with the establishment of English Partnerships in 1993, the approach to regeneration stressed a strategic, long-term, balanced and partnership-led framework that contrasted with the deregulated, targeted and property-led approach of the 1980s (Tallon, 2010). Nevertheless, the competitive element of funding under City Challenge and the SRB replaced assessments of need as an allocating mechanism for resources and sought, instead, to emphasise and reward innovation and motivation – echoing the preferred Thatcherite market-based mechanisms.

Although there were some significant changes to planning and its institutional context during the period 1990–97, it would not be unfair to characterise the period as the 'lost years' in planning. Such a missed opportunity could not be blamed entirely on government policy, as planners and the planning profession largely ignored the opportunity for strengthening the link between planning and the growing environmental movement and recast it in the role of environmental stewardship. However, while distinct from the combative approach tone of the 1980s, the twin concerns of increased centralisation and competition/market orientation of the 1990s fitted within the New Right paradigm (see Table 1.1).

The lack of willingness on the part of planning to seize the opportunity of environmental concern, combined with the increased significance of central government and the relegation of local government to a service provider, involved with processing proposals for development quickly, did little to enhance the profile of

TABLE 1.1 Characterising the New Right's approach to planning

Period	Approach	Means (in order)	Ends	Examples of change
1979–1988/9	**'Project led'** Reduce strategic role of planning and make decisions on their merits Development led	1 Market orientation 2 Rule of Law 3 Centralisation	Deregulation of controls (release pent-up demands through targeting supply side constraints) Make system more 'transparent' by reducing local planning authorities' discretion Do not alienate conservation-minded voters	Urban Development Corporations Enterprise Zones Simplified Planning Zones Circular 22/80 White Paper *Lifting the Burden*
1988/9–1997	**'Plan led'** Reintroduce a (modified) element of strategic role for planning that takes strong central guidance into account (Illusion of) locally led planning to reduce responsibility of secretary of state	1 Centralisation 2 Rule of law 3 Market orientation	Control planning through strengthened central mechanisms (Grudgingly) introduce environmental concerns and design criteria	Elevation of importance of development plans Revised PPG series Rio commitments Section 54A *Our Common Inheritance*

Source: Allmendinger and Thomas (1998: 251).

the profession. This is not to say that planning was merely a passive recipient of change. On the contrary, there was resistance to and 'creative interpretation' of many of the proposals that were introduced through the 1980s and 1990s, particularly in the field of planning (Brindley *et al.*, 1996; Allmendinger and Thomas, 1998). This raised a fundamental issue for any assessments of change in planning under the New Right. While it is possible to identify a coherent and determined policy direction, we should be wary of taking as read the impacts at other levels of government. In the conclusion of one assessment: 'The Thatcher governments may have had more radical objectives than previous governments, but they were probably no better at achieving those objectives' (Marsh and Rhodes, 1992: 170). One of the outcomes of the realisation that even a determined and uncompromising government could not always effect change was a re-examination of the 'Westminster model' of central–local relations and its replacement with a more nuanced, institutionalist understanding around networked governance that emphasised coexistence and interaction between different arms and scales of government (Marsh and Rhodes, 1992).

The experiences of the New Right also highlight a further issue in analyses of change that depends upon political ideology as the benchmark: 'Thatcherism was an evolving process which changed significantly over the eighteen years the Conservatives were in office, in the process becoming more coherent and more radical, ultimately culminating in the reforms which characterised the party's third and fourth terms' (Kerr and Marsh, 1999: 187). Even a cursory look at the history of Labour in power demonstrates a similar evolution of policy and approach. Nevertheless, as Table 1.1 makes clear, evolution and change during the New Right era involved a rearrangement and reprioritisation of themes and tenets rather than a wholesale replacement of ideology itself. While it is possible to identify distinct periods within the New Labour era in relation to planning, these too exhibited common foundations (see Chapter 2). Also echoing the experiences of the New Right, such common foundations or tenets were, at best, in tension, and, at worst, in contradiction.

At the cusp of transition from the Major government to the Blair administration in 1997, planning found itself in a broadly familiar position.

- Despite some experimentation, UK planning remained fixed to the separation of plan and permission, as it had been since the 1947 Town and Country Planning Act. The gap between the two had been bridged through the introduction of a 'presumption in favour of development', which placed the onus upon local planning authorities to demonstrate why a proposal should not proceed. This approach had been a half-hearted attempt to increase speed and certainty within the system, much in the same way that zoning-based approaches did in other jurisdictions. The outcome was that, in an under-resourced system during a development boom, planning was characterised by delays and uncertainty.
- Regardless of much of the rhetoric of the time, planning remained based upon

the interpretation of national policy guidance within local contexts. Local and professional discretion persisted at the heart of the system in spite of the expansion of central planning guidance and the increase in prescription and detail.
- The moves towards creating a more market-orientated and supportive planning system had largely been thwarted by resistance from more conservative-minded MPs, the opposition of the rural electorate and a misunderstanding of land and property markets, and the relation between scarcity and profit. Planning had always been and remained largely market supportive (McAuslan, 1981) and, at the end of the New Right era, continued to be based around a 'predict and provide' approach to the allocation of land for development.
- Despite some attempts to narrow the range of issues or material considerations, planners followed the broad scope provided in case law that a material consideration was 'any consideration which relates to the use and development of land' (*Stringer* v. *Minister of Housing and Local Government* [1970] 1 WLR 1281). In other words, they could justify incorporating a wide range of issues and local matters into decision-making regardless of the shift in advice and the presumption in favour of development. An applicant needed to appeal to test the validity of the scope of the issues that a local planning authority considered material, which was a lengthy and costly process.
- While some services within local government had been hived off to the private sector or reorganised around a client–contractor split, planning remained largely a local authority function, even though the use of consultants to provide specialist services had increased.
- Initiatives such as Urban Development Corporations and Enterprise Zones had been rhetorically cast as radical, although they owed as much to earlier models of New Town Development Corporations as New Right thinking. In practice, such special planning arrangements never amounted to anything more than experimentation.

While some asked: Whatever happened to planning? (Ambrose, 1986), it was clear that planning, unlike other areas of state activity or public policy, had not been radically altered.

The experiences of planning during the 1980s and 1990s raised a number of questions and lessons for the incoming Labour government, not least around the resilience of professional and local autonomy and the co-dependence of the centre and local planning authorities and regimes in effecting change. Where there had been demonstrable changes in planning, they had tended to exclude or bypass local authorities. The changes to the Use Classes Order or the imposition of Urban Development Corporations or Enterprise Zones did not involve or require the cooperation of local planning authorities and, consequently, were implemented more or less as envisaged and were more successful in achieving the desired outcome. Other initiatives that were dependent upon local and/or professional discretion were far less effective in bringing about change and, at times, were used for a variety of ends that were widely at variance with the original intentions (Allmendinger and Thomas, 1998). Where there had been more positive moves towards providing a central role for planning in the emerging spheres of climate change and environmental stewardship, these had been largely ignored by planners and local authorities too.

Approaching New Labour and planning

The overview of planning under the New Right points to the importance of continuity and stability in any assessment of change. The experiences of the New Right and planning also highlight the difficulties in evaluating change. Evaluations need a counter-factual against which to compare change. Focusing upon specific elements of Labour's spatial policy or planning governance could have a reasonable stab at identifying what would have occurred in the absence of deliberate change – for example, performance targets in development control (see Chapter 6) or the new system of development planning (see Chapter 5). However, there were significant areas of change in planning under Labour that were either unintentional or contradictory. For example, just as the government began to focus its attention upon housing delivery, the credit crunch and then recession brought development to a virtual standstill. It was almost universally accepted (except by ministers) that the proposed Planning Gain Supplement would have reduced development activity at a time when the government was seeking an increase. A further problem in an era of complex, issue-based governance is the possibility of unpicking the impact of discrete changes from among what are inevitably 'packages' of multiple policy streams and initiatives. For example, is it possible to separate and evaluate the objectives of the new system of development planning that was introduced from 2004 from the effect upon its implementation of the widely recognised shortage of planners? Could the reasons for the lack of progress on plan preparation be identified as systemic, resource based or both?

The second prerequisite in evaluation is a theory of change or the necessity to indentify a clear intention to change. As with the New Right, there was not an agreed underpinning to policy under Labour. Elements within Labour, old and new, privileged and emphasised different aspects and orientations, depending upon the issue or point in time, from a spectrum spanning more traditional social democracy to pragmatic neoliberalism (Hay, 1999; Finlayson, 2009). The approach between government departments varied as much those between ministers within departments. The Treasury had a sceptical view of the role of planning and was concerned with its impact upon competitiveness and the economy. In its view, planning was a supply-side constraint upon competitiveness and economic growth. The department that had responsibility for planning, depending upon the minister, at times concurred with this view, particularly between 1997 and 2002, or took a very different, contrary perspective. This lent uncertainty and ambiguity over the actual direction of policy, strengthening the hand of those charged with rolling out change in being able to 'pick and choose' which policy narrative to implement.

A further issue that arises from the experience of the New Right and planning is that change is unusual if not the exception. Within government, institutional inertia, culture and path dependency all play an important role in mediating, thwarting and curtailing the intentions of Parliament and the executive. The institutionalist perspective that frames this book (Chapter 3) points to a range of sources of stability and change and highlights that 'institutional change is never a technical matter, because any challenge to existing institutional settlements is likely to be met by resistance' (Lowndes, 2005: 294). This highlights the importance of exploring not only what Labour intended but how change was mediated through existing professional and governance arrangements. Similarly, while the prefix 'New' was used in relation to the Conservative administrations throughout the 1980s and 1990s as well as the subsequent Labour governments, we should be wary of the lure of the 'new' and the invitation to underplay continuities. There was an identifiable trajectory in the themes of and approaches to planning from the New Right to the New Left, with Labour actually building and expanding upon some of the important tenets of New Right thinking. For example, under New Labour the replacement of Section 54A with Section 38(6) of the 2004 Act saw a further strengthening of the significance of the development plan, with the change from 'shall have regard' to 'must have regard'. The 'test of soundness' for development plans introduced as part of the 2004 changes required that the plan be in conformity with national planning policy, further strengthening the centralisation of policy and minimising local discretion. The continuities and evolutionary nature of change in planning were more significant than is generally acknowledged. For example, the Citizens' Charter and Compulsory Competitive Tendering morphed into the Best Value regime introduced by Labour in 1999. While Best Value sought to shift the focus of the Citizens' Charter and Compulsory Competitive Tendering to quality as well as quantity, planners and others who were subjected to Best Value Performance Indicators and the Planning Delivery Grant may disagree (see Chapter 6). In urban policy the Single Regeneration Budget and the underlying philosophy of competition introduced under Major continued under Labour, though it was supplemented by other initiatives and schemes.

Before embarking upon assessing the approach and impact of New Labour and planning, there are a couple of issues that need to be addressed. First, while I discuss the nature of New Labour in Chapter 2, it is worth stating at the outset that I use the term 'New' Labour not because I accept that it represents a departure from 'Old' Labour but because it was the widely used moniker which the party chose to describe itself. As I argue in Chapter 2, there are a variety of positions on the extent to which New Labour was a continuation of Thatcherism, a pragmatic, ideology-free zone or a form of social democracy for New Times. Rather than taking an a priori position, one of my objectives in this book is to come to a view on the nature of New Labour from an empirical analysis of its approach and policies for planning. Second, this book centres on England. A history and evaluation of planning *per se* during the period 1997–2010 would clearly need to include Scotland, Wales and Northern Ireland. However, the book's focus upon planning *and* New Labour naturally limits its scope to England, as for parts of period in question Labour was either in coalition government or in opposition in the devolved administrations.

Labour inherited a planning system that had changed little since 1979. The party entered government with the intention of changing and modernising, particularly in the public services, but without a clear idea of how to go about this. The chapters that follow provide a necessarily selective and subjective illustration of changes to planning during the period 1997–2010. A different focus may well provide a different picture and justify different conclusions, but the focus was driven by the need to provide a cross-section of change covering different elements of planning.

The structure of the book is as follows. Chapter 2 explores the nature of New Labour and its approach to planning, concentrating in particular on the identification of distinct periods of change between 1997 and 2010 that can be allied to specific attitudes towards planning. In Chapter 3 I set out a framework of understanding of change and governance. The significant discretion afforded planning and localities as well as the move towards more networked, multi-scalar governance points to an approach that accounts for the complexity, evolution, professional autonomy and variability in local practices. Chapters 4 to 7 constitute the four evaluative chapters of change to planning. Chapter 8 comes to some conclusions, reflects upon the framework of understanding in Chapter 3, and presents a way of comprehending both the nature of change under Labour and how the system is now evolving under the coalition government.

2
NEW LABOUR AND PLANNING

Introduction

This book is about changes to planning during an era when there was an attempt to 'modernise' it. As I discussed in Chapter 1, this is not the first time that a government has sought to recast planning in order to achieve its objectives. Yet, analysis of such change is not a simple matter. It is tempting to harmonise approaches and periods, smoothing over complex and nuanced objectives, conflating intended and unintended change, confusing rhetoric with action, and falling into the trap of confirmation bias to support a particular view. Did New Labour succeed where the New Right failed in having coherent objectives for reform and successfully implementing its agenda for planning modernisation? According to the then planning minister, Lord Falconer, in a speech to the CBI on 26 November 2001, the battle to reform planning had been fought and won: 'The time is long overdue for a wide ranging reform of the planning system ... Planning is not serving business well – but neither is it serving the community.' And Baroness Andrews, parliamentary under-secretary of state, in a speech to the Planning Inspectors Conference on 29 March 2006, said: 'We have, together with other colleagues in other parts of the planning world, turned around a service that was seen as tired, out of date, and sclerotic.'

These two statements, both from ministers with responsibility for planning, were made less than five years apart. There is evidence to support the idea of a 'planning renaissance', as I discussed in Chapter 1. According to some, particularly the professional bodies, planning was reborn under New Labour. In its guise as the new *spatial* planning, it was the key nexus in the landscape of spatial governance, coordinating, engaging and managing the processes of change in order to deliver sustainable development (e.g. Nadin, 2007; UCL and Deloitte, 2007; Davoudi and Strange, 2009).

Yet there were also clear signs of frustration at recalcitrant practices and views within planning, a lack of delivery and recidivist tendencies. This led to disenchantment with planning within government about the possibility of evolutionary reform. With the onset of the credit crunch, and then recession and its impact upon development from 2008, dissatisfaction with planning grew into commitment and determination for further change. The 2008 Planning Act heralded the start of yet another era of 'improvement' and created the Infrastructure Planning Commission to deal with strategic infrastructure proposals, as well as shifting responsibility for preparing Regional Spatial Strategies to Region Development Agencies and introducing a new approach to planning obligations through the Community Infrastructure Levy. These changes could be portrayed as part of an evolution of the system, though the 2007 Planning White Paper (DCLG, 2007a) highlighted the government's frustration at the (continued) length of time taken to prepare development plans. While the approach and some of the objectives for planning differ significantly from those of the Thatcher governments, it would seem that New Labour had its own 'implementation deficit'. Further radical reform was heralded by the Conservative Party, which in opposition prepared its own critique and 'plans for planning' based upon the view that the new regional level of development plans was undemocratic, that planning needed to be much more locally driven, and that local communities should be incentivised to accept more development rather than being 'forced' (Conservative Party, 2009, 2010): 'The planning system is vital for a strong economy, for an attractive and sustainable environment, and for a successful democracy. At present, the planning system in England achieves none of these goals. It is broken' (Conservative Party, 2010: 3).

As in the Thatcher period, there were multiple and irreconcilable views of the objectives, underlying vision and impact of changes to planning under New Labour. This chapter explores the government's intentions for planning in opposition and power. Despite devolution, the UK remains a largely unitary state where the intentions and policies of the Westminster governing party still dominate. This is not to say that any legislation or policy is then implemented directly or uniformly – the 'Westminster' model of government, which is based upon a clear separation of powers and roles between the various tiers, has given way to a more nuanced and empirically based understanding of governance (Stoker, 2000; Bevir and Rhodes, 2006) and the emergence of neo-institutional frameworks of understanding on the contextualisation of and influence upon action and outcomes (March and Olsen, 1984; DiMaggio and Powell, 1991; Lowndes, 1996; Bevir, 2005). However, in areas such as planning, where there is a legal framework and requirement to 'trickle down' forms of policy through multi-scalar institutions and processes, national planning policy remains an overarching framework within which regional and local development plans, along with individual planning decisions, should conform (ODPM, 2005a). The starting point in any evaluation of change remains at the national level. This is not to say that the local practices of planning are irrelevant, as one of the arguments of this book is the significance of planners 'on the ground' (or 'street-level bureaucrats', as Lipsky (1980) termed them) in creatively interpreting, transforming and twisting policy to meet needs and circumstances other than those envisaged by government. In addition to local-level policy interpretation there are examples of 'bottom-up' policy emerging and influencing national policy, particularly in the area of renewable energy (Wilson, 2009). I develop this framework and understanding more in the next chapter. The rest of this chapter is seeks an understanding of New Labour and its approach to planning.

New Labour

It is becoming common to evaluate change during the New Labour era as involving tensions between what were seemingly competing and irreconcilable objectives common in all social democratic parties (see, for example, Bevir, 2005; Stoker, 2004; Stoker and Wilson, 2004; Lowndes and Wilson, 2003; Taylor, 2009). Planning is no exception to this, and a range of commentators – for example, Colomb (2007), Inch (2009), and Ellis (2007) – have highlighted the ambiguities and contradictions in Labour's 'urban renaissance' rhetoric. Analyses of this agenda can focus, for instance, upon the efficacy and implications of the different tools and vehicles used (e.g. partnerships, area-based initiatives and the financial incentives) or the desire to improve local services, reduce crime and anti-social behaviour and create attractive places for the middle classes to recolonise the city. Equally legitimately, analyses could explore housing market renewal or the administrative and institutional complexity of growth areas such as the Thames Gateway. The obvious consequence of this multiplicity of themes and objectives is that there is little consensus on the coherence, significance and impact of change during the Labour era. One way in which to address this is to step back and pose more fundamental questions around the nature of Labour and how its essence evolved and influenced policy. In short, *what* New Labour, *which* New Labour and *when* New Labour?

What New Labour?

Some analyses seek to understand the nature of New Labour by attempting to highlight and explore the differing component influences and then gauge whether this amounts to a coherent programme or philosophy (e.g. Gamble and Wright, 1999; Coates, 2005; Bevir, 2000). But, as Finlayson (2009) points out, over a relatively long period in power governments can develop policy, legislation and initiatives and explain them in variety of different ways to different audiences. This makes assessing any kind of coherence or link to ideological underpinning less than easy. New Labour had many faces and, depending upon the issue and perspective of the analyst, was a continuation of neoliberalism (Hay, 1999; Callinicos, 2001), part of a traditional, northern European form of social democracy (Driver and Martell, 2006), a unique fusion of social justice, economic efficiency and personal responsibility (Giddens, 1998; Le Grand, 1998; Wilks-Heeg, 2009) or a pragmatic, managerialist approach that defied ideology and simple classification (Finlayson, 2009).

Even when assessments of change are taken into account there is still widespread disagreement. For example, drawing upon Lowndes and Wilson (2003), we can identify five main schools of thought in relation to change in local government under New Labour. First, there is the argument that Labour initially suffered from a combination of naivety, enthusiasm and pent-up demand to 'make a difference' after being in opposition for eighteen years. Such 'newness' led to initial high expectations and a 'hotch potch' of initiatives and programmes (Lowndes, 1996; Stoker, 2000). Second, there are those who argue that behind the apparent eclecticism of initiatives is a programme aimed at centralising control and enforcing change and modernisation. Pragmatism on means diverts attention away from the distributional consequences of 'taking sides' (Davies, 2001; Geddes and Martin, 2000). The third school of thought is less cynical than the centralisation argument above and maintains that New Labour undertook an approach based on 'principled pragmatism'. The apparently contradictory and eclectic range of initiatives can make more sense when viewed as a conscious attempt to reject traditional solutions from either the right or the left of the political spectrum. The 'what works' approach eschews dogmatic responses and existing institutional arrangements in favour of evidence-based policy and mechanisms (Finlayson, 1999).

Fourth, the range of approaches can best be understood as a response to the emergence of multi-scalar, networked governance, privatisation and the fragmentation of the public sector. Problems or issues that the state seeks to tackle have also become more complex and interrelated. Attempts to improve health rather than simply health care, for example, required an eclectic approach and institutional infrastructure recognising that 'messy problems' require 'messy solutions' (Rhodes, 1997). Finally, there is the claim that New Labour tackled the modernisation of the public sector by attempting a range of approaches and methods in order to test different solutions. This 'policy laboratory' method is echoed in the underlying thinking behind devolution, though the actual scope for autonomous action in the devolved administrations was tightly circumscribed (Haughton *et al.*, 2010).

The point is that the different accounts of change above accurately reflect the pragmatism and eclecticism of New Labour: the expectation that any government has a coherent and identifiable ideology is outdated and a residue of an earlier era. One of the criticisms of understandings of New Labour is that we should not expect to find consistency. Bevir, for example, attacks the 'bland, complacent, evasive and mistaken' analyses that attempt to provide a coherent, uncritical 'story' by linking rhetoric and intention to change 'on the ground' (2005: preface). For planning this is an important point, given the plethora of approaches and policies over the Labour period that appear to come from a variety of origins. Some of these predate the Labour era (e.g. the commitment to green belts and urban containment), some are clearly ideologically driven (e.g. public-sector 'modernisation'), some are pragmatic responses (e.g. the post-2004 system of development planning) and others appear to have emerged as consequences of other events (e.g.

the role of planning in helping address terrorism). It is also possible to point to a range of policies that, at the very least, sit uneasily together and, at worst, contradict each other. For example, the commitment to an urban renaissance and regeneration could be undermined by recurrent proposals to relax retail policy for out-of-town developments. Urban regeneration as a policy theme includes a wide variety of more detailed policy instruments that can broadly be traced back to progressive social democratic concerns with social inclusion and justice; greater out-of-town retail competition and choice, on the other hand, favour a narrower demographic while also helping undermine regeneration in a variety of ways.

Ideological labels are monolithic and reify some aspects of change over others. As Finlayson (2009) rightly points out, if we look for evidence of neoliberalism within the policies of New Labour we can find it. Likewise, if we look for evidence of social inclusion and localism we can also highlight numerous examples. If we move beyond New Labour as a whole and begin to break down the 1997–2010 era we can unpick some of the generalisations and categorisations and identify distinct periods when certain ideas were more dominant than others. I attempt this further on. Nevertheless, we must take seriously the idea that the old rules and classifications no longer apply. In an era of networked governance across multiple scales, embedded within a global economy and institutions, the room for ideology is severely diminished. Powell (2000: 53) gets it about right with his conclusion that the big idea of New Labour was that there was no 'big idea':

> New Labour has not forged a hybrid out of social democracy and neo-liberalism, and nor has it split the difference between them. It is created not an ideology but an orientation – a way in which problems are identified and solutions formulated out of elements culled from a variety of intellectual and technocratic traditions.
>
> *(Finlayson, 2009: 17)*

That there is no coherence should come as no surprise. The coherence of the New Right was no such thing – it was itself an amalgam of various themes based around the free market and the strong state (Gamble, 1988) and evolved in response to both internal and external influences. During the 1980s the approach on any one particular issue would vary (many areas of policy did not easily map onto approaches that could be 'rolled out' from underlying themes) according to the minister concerned, public opinion and the electoral impacts. These 'day-to-day factors' were mediated in a competitive arena of themes – free market or strong-state conservative – that would help shape the approach.

Which New Labour?

Given the above, a slightly different way of trying to understand the 1997–2010 era that has more resonance for planning is considering *which* New Labour?. It is clear that sources of change for planning during the New Labour administrations originated from a range of sources within (and outwith) government, including other government departments. Wilson (2009), for example, highlights the importance of the Treasury in driving forward policy agendas and the impact this has had upon planning controls. It will be clear to anyone involved in planning during the period that the two Barker inquiries (2004, 2006a), while ostensibly multi-departmental, were, nevertheless, Treasury initiated and driven. Differences between departments and the dominance of one view was, in part, aided by the emphasis under Labour upon joined-up, issue-based government rather than more traditional policy sectors or silos:

> Many of the biggest challenges facing Government do not fit easily into traditional Whitehall structures. Tackling drug addiction, modernising the criminal justice system, encouraging sustainable development, or turning around run-down areas all require a wide range of departments and agencies to work together. And we need better coordination and more teamwork right across government if, for example, we are to meet the skills and educational challenges of the new century or achieve our aim of eliminating child poverty within twenty years.
> *(Performance and Innovation Unit, 2000: prime minister's foreword)*

This 'issue-based' approach allowed the more dominant departments within government to influence other areas of policy more readily, an opportunity that allowed the Treasury, in particular, to strengthen its authority. This dominance depended, in part, upon the individuals involved, though the significance of individuals also helped shape a particular approach within a department. For example, the direction and pace of proposed changes to planning clearly differed between Stephen Byers and Lord Falconer, as ministers with responsibility between 2001 and 2002, and John Prescott, from 2002 to 2006. The former two, as secretary of state for transport, local government and the regions and minister for housing, planning and regeneration, respectively, pursued a clear, business-orientated programme of planning reforms, while Prescott introduced a greater emphasis upon planning as a nexus for a range of concerns around, *inter alia*, climate change, social justice and regional economic development.

When New Labour?

The final dimension in understanding the nature of New Labour concerns the temporal and how attitudes and policies evolved and changed through time. Any political party in power for a long period will evolve through experiences of policy and through external factors, including shifts in public opinion. Breaking down such long periods in order better to understand different emphases and approaches is not uncommon. For example, Wilks-Heeg (2009) identifies distinct periods in relation to changes in local government under New Labour. The early era, roughly coinciding with Labour's first term, was dominated by attempts to make local government 'modernise' while also seeking to improve local democracy (ibid., 2009: 28). Later approaches were concerned more with targets and performance. However, it is worth stressing that the democratic renewal stream did not disappear during later phases but was merely eclipsed by a more dominant concern.

From even a superficial assessment it is clear that there were identifiable and distinctive eras in Labour's approach to planning, regardless of the change in leadership and prime minister (Allmendinger and Tewdwr-Jones, 2009). Particularly notable ruptures in approach occurred in 2001, 2003 and 2008. The allure of periodisation of any era needs to be balanced against the inevitable dangers of bringing together multiple threads of evolving and overlapping policy, time lags and the need to generalise across space and scale. It is also necessary to recognise that many of the important changes to planning occurred outwith 'planning'– for instance, local government modernisation.

In terms of what New Labour?, which New Labour? and where New Labour?, what can we conclude? While views inevitably vary, the overall picture is that, with regard to planning and related areas, it had a 'light' social democratic ideological orientation that combined liberal, market solutions with a more pragmatic attitude towards the means by which policy would be rolled out:

> New Labour has brought a greater concern with choice to the socialist ideal of social justice, a greater concern with duty and responsibility to the socialist ideal of citizenship, and a greater concern with competition and manipulation to the socialist ideal of community.
>
> *(Bevir, 2000: 298)*

Increasingly, however, such early characterisations became muddied or less telling as different ministers introduced policies or as events (such as the recession) led to the need to introduce policies that did not easily fit, if at all, within such understandings. This 'drift' was facilitated by the use of overarching themes of change, such as 'modernisation', that provided useful and vague vehicles to encompass a plethora of evolving measures and ideas.

This is not to dismiss the idea that we can identify or pin down the nature of New Labour. Such analyses provide useful starting points. They also suffer from inevitable drawbacks, including the lack of engagement with planning within the context of local governance, which – as I go on to set out in the next chapter – provides for both interpretative discretion and implementation of such agendas. The result is a tension between national, necessarily general policy agendas and local needs and characteristics. The upshot is inevitable spatial variation in outcomes. Further, there is also an important element of 'feedback' as monitoring and evaluation of policies comes to influence their evolution. Consequently, such understandings can take us only so far, and there is a need to derive the nature of New Labour empirically and not simply from trying to read off the significance of policies and initiatives from overarching ideologies or positions.

New Labour and planning

As might be said of New Labour generally, the experiences of planning between 1997 and 2010 excite, disappoint and divide. They also exhibit similar tensions and contradictions as those within the New Labour project as a whole. As with any area of public policy there are numerous and conflicting views on the character and impacts of change:

- For some the shifts towards more spatial forms of planning have led to its becoming a sectoral 'glue' or epitome of networked governance (e.g. Albrechts, 2006a, 2006b; Healey, 2007). Others point to the lack of progress in plan-making caused by the difficulties of coordination and the lack of 'buy in' from planners and others (e.g. Newman, 2008), continued 'silo' mentality within local authorities and planners (Cuff and Smith, 2009; Centre for Cities, 2010) and a retrenchment into regulatory forms of control (CBI, 2005; Zetter, 2009). Yet others have highlighted the role of spatial planning in working against delivery in places through adding to the 'congested state', creating unachievable expectations and failing to appreciate the difficulties in multi-scalar, sectoral coordination (Allmendinger and Haughton, 2009b).
- In development control (or development management, as it was rebranded) the imposition of performance targets coupled with financial incentives has made either significant impact on the speed of application determination (DCLG, 2006e, 2007a) or little overall impact (Killian and Pretty, 2008; NAO, 2008; Ball *et al.*, 2009).
- While the government has sought to place community involvement at the heart of planning (ODPM, 2004b) there is also the claim that other changes have actively undermined these objectives (Allmendinger, 2009), emphasising the gap between rhetoric and reality in public involvement (Brownill and Carpenter, 2007).
- Although a long-term concern of successive governments (Alexander, 2007), economic development and the role of planning has been either too dominant a theme in change (Greenhalgh and Shaw, 2003; Hall, 2003) or insufficient (Simmons, 2008).
- While new regional planning arrangements are heralded by some (Marshall, 2007), they do not sufficiently emphasise sub-regional planning (Baden, 2008) and are heavily criticised and identified for abolition by any future Conservative government (Conservative Party, 2009, 2010).

The catalogue of views of the impacts, trajectories and implications of change could go on and merely reflects the highly political nature of the activity of planning. We should expect this diverse range of views to exist. Although the above list is by no means exhaustive, the general point is that there is evidence of, or at least claims to support, change and continuity, success and failure in the 'modernisation' of planning.

There are inevitably a variety of schools of thought on planning and change under Labour. Some, such as Taylor (2009), claim that New Labour suffered from the same tensions in its ideology that befell other social democratic governments. For Taylor, disappointment over the impact of Labour's modernisation of planning derives from the continued need to persuade capital to implement government objectives. So, while Labour could claim to be more positive about planning and desire greater involvement, it ultimately had to be market sensitive to ensure objectives were achieved. Others, such as Wilson (2009), argue that planning under Labour moved from being a rather insular and inward-looking regulatory process to being one that embraced and had to come to terms with global concerns around the environment. Labour's strong centralising tendencies were necessarily overcome by the need to tackle climate change through planning at different scales and tiers. Under this view there may have been an ideology driving change but, if so, it was lost by a succession of legal challenges, local action and competing views of sustainability within and outside of government. For Wilson, changes to planning were evolutionary, pragmatic and 'work in progress'.

There is also a danger, as hinted at by Kunzmann (2009), that such assessments were disproportionate: planning was and remained a relatively minor element of government policy. As he goes on to point out, continuity between administrations in a small area of government policy is probably the norm, and he would find it difficult, for example, to distinguish between the planning policies of the Social and Christian Democrats in Germany. A related position which argues for a longer perspective on change points to the similarities between the fundamentals of the post-1947 planning system and the 'modernisation' under New Labour (Prior, 2005). The *longue durée* of reform is a useful reminder of the intractable and tension-laden nature of the state's role and its reflection in the nature of planning. I cover this is more detail in the next chapter. However, it is always useful to have this perspective in mind when coming to the conclusion, as many do, that planning under New Labour is indistinguishable from the New Right approaches of the 1980s. There are continuities *and* differences between the Conservative and Labour governments, though, as I touched upon in Chapter 1, it depends very much on which New Right that Labour is compared with. The high-water mark period of deregulation, following the 1983 general election, cannot be easily compared with the 'plan-led' re-endorsement of planning from 1991, though both could be argued to fall within the 'New Right' era. Nevertheless, there is continuity through the focus on speed of decision-making and centralisation of policy and control. But there are also continuities with themes of planning from regimes before 1979, including a strong social dimension in planning policy. As well as reinforcing the dangers of periodisation of change, the 'continuities and ruptures' debate highlights how planning can change without significant legislative amendment and how some themes are consistent regardless of central intentions:

> Planning, in England at least, appears to be splitting up in many ways that suggests both continuity with the past (development control management, for example) and radical reform (spatial planning, that emphasises corporate

governance, turning back the clock towards a form of civic planning not seen since pre-1970s albeit within a fundamentally changed state).

(Tewdwr-Jones, 2009: 684)

Echoing the arguments that New Labour's approach to local government has involved a postmodern, eclectic mix of cause and effect, Inch (2009) points to the uneven, restless and pragmatic search for more effective governance and planning. The influences upon the directions of change are identified by Inch as coming from a crude 'planning inhibits development' agenda in the Treasury that was heavily influenced by the CBI. The other driver of change was an effective reinvention of planning as being a positive, significant contributor to modernisation through its ability to bring together both as process and output New Labour concerns over the economy, society and the environment.

The 'planning in new times' thesis accepts that the traditional role of the state has changed for a variety of reasons, including the acceptance by Labour of the legacy of the New Right (Allmendinger, 2006), fundamental shifts in the capitalist mode of accumulation (Prior, 2005) and an element of restlessness and evolution:

> ... planning has been, to New Labour, two things at once. It has been an object of modernization, something that must be reformed and whose personnel must come to take on their function in emerging structures of governance. It has also been a tool or instrument to be used in bringing about just that kind of modernization.
>
> (Finlayson, 2009: 18)

By comparison, Taylor's (2009) analysis says little about the activity of planning under New Labour and how it has evolved, how different approaches by government have fared and what this tells us about achievements compared with objectives. It also fails fully to understand the 'implementation problem', as he puts it, dismissing it as a consequence of inherent contradictions in any social democratic approach to planning. As Marsh and Rhodes (1992) and Allmendinger and Thomas (1998) argue, bringing about change relies in large part upon how it is approached, and there can be more or less effective implementation.

It is clear that attitudes and policy evolved under Labour and that, despite the dangers in providing a false coherence, as discussed above, periodisation of centrally driven change in planning is a necessary if not essential component of understanding the Labour era. Mindful of the caveats, and that different emphases will produce varying results, there were four distinct periods in New Labour's approach to planning.

1995–2000: Continuity and devolution

Allmendinger and Tewdwr-Jones (2000, 2009), Inch (2009) and Wilson (2009) all argue that the early period of New Labour was characterised by a broad acceptance of the legacy, tools and objectives of the New Right. This period witnessed both a naivety about the way in which change could be effected and a lack of any clear ambition or direction away from the trajectory established under the previous Major administrations. There is evidence that, regardless of any clear view of how planning fitted into the New Labour project, reform was not a priority, given the focus upon other sectors such as health and education. According to Keith Vaz, shadow minister for planning and regeneration:

> We decided it was not going to be possible to persuade the Chief Whip in a new Labour government or a Leader of the House of the fact that planning is a hot political issue and we thought it would not be possible to get through Parliament a long Town and Country Planning Bill.
>
> *(Vaz, 1996: 12)*

This period was characterised by the lacklustre and minimalist policy statement *Modernising Planning* (DETR, 1998) and by the continuation or modification of existing policies, particularly on housing densities and brownfield targets. *Modernising Planning* has been described as a 'shopping list' rather than a coherent strategy (Allmendinger and Tewdwr-Jones, 2000). If there is an underlying concern it is with speed and efficiency – themes that had been underpinning concerns with and changes to planning since the 1970s. Headline-grabbing initiatives, such as signing up to the European Spatial Development Perspective not long after coming to power in 1997, gave the outward appearance of momentum and direction while actually making very little practical difference. Other proposals, such as Regional Development Agencies, were modelled on the perceived success of the Welsh and Scottish Development Agencies and were concerned more with emphasising economic development and regional competitiveness than planning *per se*.

The establishment of the Urban Task Force under Richard Rogers in 1998 was welcomed by many, and it helped deflect attention away from the lack of any distinctive policy direction within government. The key recommendations (DETR, 1999), on housing density, regeneration vehicles, attention on design and the integration of transport, fed directly into *Planning Policy Guidance Note 3* published in March 2000 (DETR, 2000a) and the Urban White Paper published in the same year (DETR, 2000b). The White Paper and the Rogers report were widely welcomed as representing a new commitment to cities that went beyond purely physical dimensions of redevelopment to embrace the social and economic aspects of urban life (Healey, 2004b). Nevertheless, at its core the White Paper was also concerned with some basic issues around how to manage the increasing need for development and housing without impinging upon greenfield and greenbelt land. While planners focused upon the White Paper as evidence of a commitment to urban policy, it was, in fact, representative of the ongoing fight within government and the party around the role of the state and social justice and the importance of increasing investment in public services. It was also an early example of the much repeated model under Labour of establishing an inquiry into a subject in a search for policy discourses and responses.

Such relatively minor change during this early era can be explained by the lack of a clear vision for planning and the desire to focus upon other areas of policy. In opposition there had been a number of policy reviews and documents around planning that came to little other than emphasising the need for speed and certainty. The 'vision' for planning was that it was a means towards wider ends (Labour Party, 1996). Not only was there no 'big idea' for planning, there was also no commitment to it – only a commitment to 'what worked'. However, the vacuum of strategic thinking about the role and objectives for planning was being filled by others, most notably from within the Treasury and the Department of Trade and Industry. The discursive role of 'urban renaissance' was matched by the much cruder, deregulatory 'burdens on business' emerging from within the Treasury, which had commissioned its own inquiries and policy analysis (McKinsey Global Institute, 1998; DTI, 1999; Better Regulation Task Force, 2000). One outcome was an intergovernment *mêlée* over the impacts and purpose of planning that echoed the divergent views within the two main streams of New Right thinking.

The first era of planning under New Labour was suitably bookended by a major and largely overlooked decision that could have forced a fundamental rethink on planning regardless of government intentions. The Human Rights Act 1998 allowed challenges to UK law based on the European Convention on Human Rights (ECHR) through the UK courts. This provided a quicker and cheaper route for those wishing to challenge UK legislation for compatibility with the ECHR. A decision by the High Court in December 2000 found the planning system to be incompatible with convention rights, specifically the rights under Article 6(1) to a fair trial. The specific issue was the role of the secretary of state in determining decisions and issuing policy – in other words, being part of the executive and also having a judicial function. The government appealed the ruling and the House of Lords gave its judgement in May 2001 (*Alconbury* v. *SSETR* [2001] JPEL 291). The Lords decided that the planning system was compatible with the ECHR as, *inter alia*, there is the option of a further challenge through judicial review of the secretary of state's decision.

While the decision was into Labour's second term, there had been expectations of challenges to planning under the 1998 Act for some years previously (Parker, 2001) and the possibility that the system would come under close scrutiny, causing a 'log jam' until the compatibility of planning with the ECHR was resolved (Baker, 2000). Undoubtedly, this overshadowed thinking on reform. If the planning system had been found to be incompatible with ECHR, as the High Court had originally determined, then a far more radical overhaul of planning, perhaps introducing Third Party Rights of Appeal, would have been required (HoC, 2002). The upshot was that pressure for reform would be broadly contained within the current system. The direction of such reform was emerging in favour of those who wished to see a more focused and market-supportive role for planning.

2000–2004: market reorientation, speed and delivery

If the period 1995–2000 witnessed little interest in or attention to planning, the period 2000–2004 made up for this. With the Alconbury decision out of the way, the second term secured following the general election of June 2001, a new minister, Stephen Byers, in post, and pressure from the business community and its advocates in government to reform, there was to be a clear, market-orientated direction for planning.

Stephen Byers announced the decision over Terminal 5 at Heathrow in November 2001, eight years after the submission of the planning application following a public inquiry that lasted nearly four years. Regardless of the decision, Terminal 5 became a useful stick with which to beat planning. In the subsequent proposals for change, 'Terminal 5' was frequently mentioned as shorthand for everything that was wrong with planning and used as a large part of the justification for 'modernisation' even though the proposal was almost unique and only one of hundreds of thousands of applications dealt with every year. The December 2001 Green Paper *Planning: Delivering a Fundamental Change* (DTLR, 2001a), and its accompanying documents on compulsory purchase, major infrastructure, planning obligations and planning fees, represented a step-change in the government's attitude to planning. Stephen Byers described the proposals as the biggest shake-up of planning for more than fifty years. The current system was, he concluded, 'slow, ponderous and uncertain' (DTLR, 2001b: 1).

There is little doubt that the government's intention in the 2001 Green Paper was to reduce deliberation over plans and planning decisions through a variety of measures and thereby speed the system up and create greater certainty. In its original form, the proposed replacement for local plans – Local Development Frameworks (LDFs) – had a core set of policies and then action plans for those areas where significant change was envisaged. The contentious elements of development planning – for example, allocating housing targets between authorities – were to be partly shifted to the new Regional Spatial Strategies (RSSs), which would, it was intended, be overseen by elected regional government. Setting housing targets would be the responsibility of bodies other than local authorities and central government, thereby removing a significant contentious element of local planning. Other significant changes included proposals to determine major infrastructure against a new national policy framework, replacing negotiated and time-consuming planning obligations by a variable fee, clarifying compulsory purchase procedures to enable land assembly for regeneration and development, and the introduction of Business Planning Zones, which would grant planning permission for a range of forms of development in advance. In all it was a collection of proposals orientated towards a quicker and more efficient delivery of development. Taken alongside other changes to development control, specifically the Best Value regime which came into force in 2000 (see Chapter 6) and the investment in e-planning and the Planning Portal to help facilitate best practice, there was a coherence and direction to government policy. Nevertheless, within eighteen months the proposals on major infrastructure were substantially watered down, formal sub-regional strategies were folded into the proposed RSSs, the proposal to limit the right of representations at the plan inquiry was dropped and planning obligations were to remain negotiated. There was also a subtle shift to reform development planning and development control through incentives and targets as part of the Best Value regime.

The tone and policy content shifted significantly from the Green Paper *Fundamental Change* in 2001 (DTLR, 2001a) to the subsequent *Sustainable Communities: Delivering through Planning* in 2002 (DTLR, 2002b). How did such a change come about? The 2001 Green Paper elicited over 16,000 responses. However, the key influences were the report by the Select Committee on Transport, Local Government and the Regions on the Planning Green Paper published on 3 July 2002 (HoC, 2002) and the Royal Commission on Environmental Pollution's report entitled *Environmental Planning*, published in March 2002 (RCEP, 2002). The latter agreed that the planning system needed reform but that it should be reorientated to support environmental issues, not business interests. In other words, the commission saw a fundamentally different purpose for planning. However, the select committee report was far more scathing of the Green Paper, taking issue with just about every aspect. It argued that the main point was that radical reform was unnecessary and that a more evolutionary and pragmatic approach would achieve the government's objectives without the disruption. There was also a risk that the changes would create uncertainty, alienate an increasingly environmentally aware and vocal population, and further delay progress through the introduction of a new and unfamiliar approach to development planning. One aspect that the select committee highlighted as a surprising omission from the Green Paper was the issue of resources. It argued that planning, as a government activity, was under-resourced and understaffed and the problems that the government and others identified were not systemic but a function of a lack of investment and suitably qualified professionals.

Both the royal commission and the select committee echoed the broad sentiments of consultees to the Green Paper. The proposals in the Green Paper managed to disappoint business, professional and environmental groups. The suggested LDFs were supported by only 29 per cent of businesses, 9 per cent of environmental groups and 44 per cent of local authorities who responded (DTLR, 2002c). The proposal for a statutory regional tier of planning was more popular. However, the introduction of targets and incentives through the Best Value regime secured the most support, strengthening the argument of the select committee that a focus upon making the existing system and processes work better was likely to prove more successful than radical reform.

While the government took on board the reaction to its Green Paper, another key aspect in the new direction was the change in ministers mid-way through the review. Stephen Byers, the secretary of state for transport, local government and the regions, and Lord Falconer, minister for housing, planning and regeneration, were the two ministers who led the 2001 Green Paper following the general election in June 2001. After a reshuffle both moved on in May 2002, when the DTLR was subsumed within the Office of the Deputy Prime Minister. Both Byers and Falconer were lawyers and saw the means to achieve change in planning as being primarily legislative. It was the deputy prime minister, John Prescott, on the other hand, who announced the more balanced approach on 18 July 2002, including a £350m investment in planning through the Planning Delivery Grant (PDG). Prescott took the proposals in the 2001 Green Paper and gave them a new set of objectives. Thus, the period 2000–2004 culminated in the 2004 Planning and Compulsory Purchase Act, based around the headline proposals but not the spirit of the 2001 Green Paper. The final outcome was still a new system of development planning, though not as originally envisaged. Subtle reorientations of the relationship between development plans and development control through the changes in wording to what was originally Section 54A of the 1990 Act to Section 38(6) of the 2004 Act sought to strengthen the role of the development plan, reduce discretion at the local level and create greater certainty. Incentives to speed up development control through the PDG would increase the speed of decision-making (see Chapter 6).

The shift from revolution to evolution did not play well with everyone. The CBI had published a briefing note on planning in July 2001, claiming that businesses were 'deeply frustrated by a system that fails to make those decisions consistently in a rational, speedy and user-friendly way, to help deliver genuinely sustainable development' (CBI, 2001: 1). This was followed up with evidence to the inquiry into the Planning Green Paper by the House of Commons Transport, Local Government and the Regions Committee with the view that, 'In every respect, in every survey we conduct, every business we talk to ... planning is always at the top of the agenda as a fetter on the productivity enhancement and the job creation in British business' (HoC TLGR, 2002: HC476-III). Further attacks on planning came from the Home Builders Federation, which called the system overly complex and inconsistent: 'Plans are too long and inflexible and preparation timetables are too lengthy' (Memorandum to the Transport, Local Government and the Regions Committee, Planning Green Paper, Thirteenth Report of Session, 2001–02, PGP 17). The British Property Federation added that the current system was failing everyone, and was 'overly plan-hierarchical, bureaucratic, unnecessarily complex, slow and inconsistent' (ibid., PGP 47).

While the government could argue that the new system of development planning would address such concerns, the thrust of these views chimed with the concerns of the Treasury, which had become increasingly interested in the role of planning in competitiveness and, in particular, the impact upon economic growth and competitiveness of housing affordability. In 2003 it had published its assessment of the five tests that would determine the UK's readiness to enter European Monetary Union, a prerequisite of moving to the euro. The assessment concluded that: '... the incompatibility of housing structures means that the housing market is a high risk factor to the achievement of settled and sustainable convergence' (HM Treasury, 2003: 225).

This was pointing to the role of house price inflation and property cycles in creating instability in the economy as a whole, largely as a result of the lack of responsiveness of housing supply to changes in demand. As a consequence 'the Government is undertaking further significant changes in the planning system, supply of housing and housing finance to tackle market failures, increase the responsiveness of supply to demand and reduce national and regional price volatility' (HM Treasury, 2003: 8). Clearly, such reforms were considered insufficient by the Treasury, which announced the establishment of the Barker review of housing on 9 April 2003, while the 2004 Act was still progressing through Parliament. The Barker review was tasked, *inter alia*, with looking into the role of the planning system in housing supply. While this was ostensibly an interdepartmental review, there was obvious suspicion, if not resentment, that the Treasury had extended its remit to housing policy. In his evidence to ODPM Housing, Planning, Local Government and the Regions Committee inquiry on planning, competitiveness and productivity, on 17 December 2002, Tony McNulty, parliamentary under secretary of state to the ODPM with responsibility for housing, planning and regeneration and the minister for London, said:

> It is not for me to speak for the Treasury, but I think their interest is partly borne – which I am grateful that this inquiry is demolishing a good part of – from this wide anecdotal body of 'evidence' out there and the view that somehow the planning system has been this massive impediment to economic growth, productivity and economic development generally.
>
> *(Paragraph 246)*

The combination of the launch of the Barker review and the passing of the 2004 Act created a degree of confusion around the government's intentions for planning. The original purpose of the changes had been to focus upon a pro-business agenda of speeding up the system and creating greater certainty. Yet, while the justification for the change had moved on, the vehicles (e.g. RSSs and LDFs) had not. Were the proposals any longer suited to the new objectives and would further change be initiated following the Barker report? If the Treasury was now saying that housing markets and price stability were of such national economic significance to justify Barker, then should reform not wait until the review reported? This confusion filtered through to the debates on the Bill in Parliament. Geoffrey Clifton-Brown, a member of the House of Commons Standing Committee G, stated:

> I consider this a bad Bill. An increasing number of practitioners who come to see me also think so. The planning system in this country has evolved. It has been built on hundreds of thousands of court cases. Why on earth did Lord Falconer come up with the Green Paper? We are tearing it all up and starting again. I think that we shall rue the day when we tore the whole system up. Why not instead build on the existing system?
>
> *(Hansard, 21 January 2003, paragraph 258)*

2004–2007: sustainable communities through spatial planning

The hyperactive and eclectic period from 2000 provided a legislative basis for development planning and a range of possible purposes to which the system could be put. What was clear was that reform and change would not be a one-off affair and that both the ultimate direction and the underlying purpose were in flux. To some this was an opportunity. With encouragement from the Royal Town Planning Institute, the government was sold the notion that planning was potentially far more than a 'command and control', regulatory mechanism (Allmendinger and Haughton, 2009b): planners had the skills, knowledge and position to reconcile a wide range of concerns and issues, from economic development and globalisation to climate change, affordable housing and regeneration (RTPI, 2007). According to this view, planning had been hamstrung by the New Right's deregulatory concerns, which restricted its purpose to one of bureaucratic, land-use regulation. The new planning – rebranded 'spatial planning' – would be a form of Third Way spatial governance:

> Intellectually, spatial planning was presented by a loose coalition of interests as part of the solution to the problems of fragmented governance in an era of economic globalisation and the need to address climate change. In practice, however, the emergence of a 'spatial planning' discourse was largely a reaction to New Labour's pragmatism. There was a mismatch between what Labour wanted from planning and what was on offer. The result was a realignment of planning and its rebranding to bring it in line with the New Labour project.
>
> *(Allmendinger and Haughton, 2009b: 2546)*

As the planning community began to try and grapple with the new system of development plans, it also had to come to terms with undertaking it in a 'spatial' way and to try to understand how to meet the new statutory requirement to create sustainable development. This period coincided with unprecedented levels of development activity and a quickening in the pace of local government 'modernisation'. 'Clarity' was provided by the reissuing and substantial expansion of national planning guidance, which presented the objective of creating sustainable communities as being uncontroversial and unproblematic. At the same time, the roll-out of performance targets and monitoring linked to incentives in development control naturally led local authorities to focus on applications rather than development planning.

2007–2010: delivery, refocus and scaling back

The inevitable outcome of the confusion of purpose, the introduction of a new system of development planning and the linking of resources to development control was a period when planning reverted to a 'project-led' approach reminiscent of the later 1980s. This was largely because local planning authorities struggled to introduce the new system through a lack of resources and the complexity of the process, particularly around the early approach to the 'test of soundness' (Zetter, 2009; Watson, 2009; Watson and Crook, 2009). Existing plans that had been 'saved' became increasingly out of date as time passed and as new central guidance emerged. Against a backdrop of increasing numbers of planning applications, this effectively meant that decisions were being taken against a wide backdrop of considerations – though not, as intended in the 2004 Act, the LDF.

A number of evaluations of the post-2004 approach echoed the growing frustration within government and business interests over the lack of progress on implementing the 2004 changes to development plans (DCLG, 2008b; RTPI, 2007) and speeding up the development control (Allmendinger, 2009). Such dissatisfaction accompanied a realisation that the government had created a confusing and confused system based upon multiple and potentially conflicting objectives (DCLG, 2007a) against a backdrop of growing resistance to new development at the local level (Saint Consulting, 2007, 2009). The flurry of proposals and actions that followed this growing realisation between mid-2007 and April 2010, also driven, in part, by the recession, were dominated by refocusing the system on a clearer, narrower set of market-supportive objectives. Proposals emerged to speed up the planning system through streamlining the consultation process (DCLG, 2010c), improving certainty by reducing discretion in negotiating planning obligations (DCLG, 2009e), linking development planning better to development control (DCLG, 2010b), making plan preparation quicker by reducing the time taken to adopt LDFs (DCLG, 2007g), introducing the Infrastructure Planning Commission (DCLG, 2009f), widening the scope of permitted development (DCLG, 2009g) and making the system more proportionate by reducing supporting information requirements (DCLG, 2010d). These changes were against a backdrop of increasing criticism of the planning system and the post-2004 reforms (e.g. Cuff and Smith, 2009; Conservative Party, 2010; Centre for Cities, 2010; HBF, 2010; British Chambers of Commerce, 2009).

Attitudes towards planning

Such periodisation is never an exact science, and there will inevitably be ragged boundaries and alternative interpretations. However, these issues do not detract from the main point, which is that there was no simple ideology–policy link under Labour. Rather, distinct periods are identifiable when certain ideas or approaches dominated. The exception is, perhaps, the 2000–2004 period, when there was confusion of purpose and lack of a dominant approach, an opportunity seized upon by a planning advocacy network to advance the interests of planning as an activity and profession. The question that then arises is: What were the range of competing discourses?

- *Consistent ends, pragmatic means*   Planning was a tool in the Third Way focus upon social democratic goals:

> It [planning] must not be a policy area on its own, it mustn't have a life of its own. It should be there to deliver the kind of things that are required by those who wish to help the economy and improve the prosperity of the nation.
>
> *(Vaz, 1996: 13)*

This attitude encompassed the simultaneous praise and criticism that characterised Labour's attitude towards planning: modernisation was a process, not a product. As David Blunkett put it: 'in the end, it is outcomes that matter' (quoted in Wilks-Heeg, 2009: 26). Both Barker inquiries explored alternative forms of regulation, and the government commissioned research on the efficacy of continental European planning systems in increasing housing supply (Ball and Allmendinger, 2006). New forms of regulation were promulgated which, in planning, took the form of a unique approach to controls over telecoms development (e.g. ODPM, 2002), the consideration of Business Planning Zones and more bespoke approaches such as the Infrastructure Planning Commission. The biggest impact of this ends-orientated discourse is the threat it posed to established interests in planning. 'Modernise or be replaced' amounted to an intimidation not only in planning but also in other areas of the public services. It was effective in unsettling established routines, practices and professional groups and was usually associated with some form of incentive or guidance to help encourage behaviour change. In development control, for example, this took the form of the Planning Delivery Grant, which rewarded local authorities that improved performance.

- *Planning as a critical nexus in the delivery of government objectives*   This was a more positive discourse on planning that emerged in the latter part of the New Labour era and sought less confrontation and more consolidation. In her speech to the Planning Inspectors Conference on 29 March 2006, Baroness Andrews, parliamentary under-secretary of state, said:

> The first thing I want to emphasise is that planning is important, not just to me personally but also to this government. It has never been more fully recognised as central to delivering economic success, environmental protection and social justice. Furthermore, and following from this, I want to stress that planning is *not* seen as a negative tool to restrict. We see its potential to design a better future, to create places where we would all want to live and to ensure that progress is inclusive rather than exclusive.

The flood of new national policy guidance from 2005 onwards sought to shift the objectives and process of planning towards spatial planning, linking this more positive role to the post-2004 development plan system. Planning became concerned with the 'management of change' and spatial governance through processes which sought to reconcile economic, social and environmental objectives through partnership and inclusion.

- *Economic competitiveness and globalisation* Not only did planning represent one of the last bastions of the 'command and control' form of regulation that underpinned the postwar approach, it also had a range of negative impacts upon economic competitiveness and growth. Labour was convinced that economic globalisation was a reality that had 'changed the rules' and that state–society relations needed to be 'modernised' (Hay, 1999). Thatcher and the New Right had 'rolled back' the state but had ignored the social consequences and underplayed the need for the state to provide for the social capital and institutional framework necessary for a free, competitive market (Harvey, 2005). However, Labour's own interpretation of the economic consequences of planning had two strands. On the one hand, planning had a positive role in economic competitiveness through providing high-quality environments and coordinating jobs and housing with infrastructure. On the other hand, it had a range of negative consequences through, for example, restricting the supply of land for housing and commerce.
- *Social inclusion and new localism* Planning as an activity or process was subject to the Labour theme of democratic renewal, with a focus upon public involvement and partnership. PPS12 (DCLG, 2008c), which provided guidance and advice on the new system of development plans within an era of 'spatial planning', had the explicit aim of engaging communities and individuals in the planning system, while the White Paper *Communities in Control: Real People, Real Power* (HM Government, 2008) sought similar ends. Through Statements of Community Involvement, which were introduced as part of the 2004 changes, local planning authorities could require developers to 'front load' consultation before applications were submitted, with the aim of allowing communities to participate in the development of a scheme rather than commenting upon it once submitted (DCLG, 2005). This focus upon inclusion came up most notably against the target-based performance culture of the Best Value regime, leading some to claim a centralisation of control and undermining of local democracy (Wilks-Heeg, 2009).
- *Climate change, adaptation and mitigation* In response to increasing criticism of the lack of a holistic response to the environment and climate change by the Royal Commission on Environmental Pollution (2002), the government began to reorientate planning policy to engage with and be a central plank in its approach to climate change from around 2003–4. Initially, the notion of sustainable communities was presented as an umbrella term that included environmental/climate concerns. Specific actions, such as the resistance to the London Borough of Merton's policy on a requirement that a proportion of electricity required by a new development should be generated on site, demonstrated a resistance to the role of planning in engaging with aspects of climate change mitigation. However, as part of the target for reducing total UK CO_2 emissions by 60 per cent come 2050, the government also set a target of zero carbon development by 2016, using a combination of the planning system and building regulations. Specific targets were linked to planning,

including reductions in greenhouse gases, increases in electricity generated from renewable sources, and increases in the area of certain wildlife habitats (Wilson, 2009).

Perhaps the most surprising element of the shift to a greater emphasis on the environment and climate change came from the CBI, which, closely following the publication of the Stern review (Stern, 2007), supported the publication of the Supplement to PPS1 (DCLG, 2006f): 'The incorporation of Planning for Climate Change as a supplement to PPS1 Delivering Sustainable Development is a clear demonstration of the importance of climate change for delivering the government's sustainable development objectives. This is logical' (CBI, 2007: paragraph 5). This support was despite the acceptance that the measures included in the draft PPS would increase costs of development. Nevertheless, as the remit from the Barker review of housing made clear, the government (or, at least, the Treasury) remained concerned that climate change in general was being considered by local authorities and others as a reason to halt development. The remit of the Barker review included the need to 'consider whether there is the appropriate balance between economic, social and environmental objectives in the English planning system' (SDC, 2006: 2). As Wilson (2009) has pointed out, there are numbers of examples where conflict between the needs of the environment and climate change have not outweighed the arguments for development based upon economic development and national competiveness.

- *Planning as networked governance in an interdependent world* The shift towards multi-scalar and sectoral, issue-based policy drawing upon partnerships between public, private and voluntary bodies needed to be secured by a form of 'governance glue' (Haughton *et al.*, 2010). Planning, through its professionals, processes and outputs, had the necessary experience and skills to enable this. Crucially, planning was one of the few areas of local government that had not been 'privatised' through Compulsory Competitive Tendering and still constituted a large, statutorily necessary core within local authorities. According to Ruth Kelly, secretary of state for communities and local government, in an oral statement to Parliament on the Planning White Paper on 21 May 2007, the opportunity for planning to constitute the 'hub' in the evolving nature of governance under Labour was obvious though not unconditional:

> today we ... face significant and growing challenges that could not have been imagined sixty years ago: from climate change, to globalisation, to energy security in an uncertain world. If we are to meet these challenges successfully, planning must be part of the solution. In its current form, it is simply not up to the task.

Such co-existing discourses at times complemented each other and at other times were, at best, in tension and, at worst, in contradiction. A major issue was the overall objective and direction for planning which those charged with devising and implementing policy had to determine. It was possible to see planning as the vehicle through which the modernisation agenda and the main tenets of New Labour would be addressed. It was also possible to see it as part of the problem – an unreconstructed impediment:

> This has made New Labour especially difficult for many public servants and their representative bodies to understand, where simultaneously they have been subject to discourses (and processes) of modernization that have cast them as part of the problem, but also hailed as potential agents in building 'modern', creative forms of governance.
>
> *(Inch, 2009: 87)*

Such discourses provided the overarching themes for planning during the Labour era, allowing an exegesis of policy to support most positions and arguments. Planning was about more housing *and* protecting the environment; it was there to create certainty and quick decisions for developers and applicants *and* to involve stakeholders in decisions about their localities. Such 'win–win–win' outcomes existed only at an abstract, rhetorical level. We cannot look simply to national-level initiatives to gauge the changes to planning under New Labour but need to see how such discourses were employed and resolved in places and what values, ultimately, persevered.

Conclusions

The multiple, evolving and conflicting nature of objectives for planning under Labour needs to be overlaid on an understanding of planning itself. Labour, like others involved in the field, tend to talk of planning in the singular when we should be distinguishing between the development plan and development control hiatus. The former is messy, deliberative, local politics in action while the latter remains largely a regulatory function (Allmendinger and Ball, 2009). Labour approached both arms of planning in different ways, incorrectly assuming that the hiatus could easily be bridged and that decisions in development control would automatically follow from the new system of development plans. This approach fell apart for three reasons. First, development plan progress was, at best, slow. Second, development control was and continues to be a highly contested and political component of planning. Finally, few, if any, of those involved were clear on what planning was there to achieve. Multiple and competing discourses created confusion which, when combined with the need to implement a new system of development planning in order to achieve 'spatial planning', inexorably led to uncertainty, misunderstanding and the opportunity to pursue different agendas at the local level.

In part, such a misunderstanding could be explained by the multiple pointers for change which came from within and outwith government. The 'contracting out' of policy formulation was to become a regular feature of Labour's approach to policy-making in general and planning in particular. Numerous successive studies or inquiries were established, among them the McKinsey report on productivity in the UK economy (1997–8), the Rogers review of urban regeneration (1998–9), the Barker review of housing supply (2003–4), the Egan review of skills for sustainable communities (2003–4), the Barker review of land-use planning (2005–6) and the Killian–Pretty review of speed in development control (2007–8). These named reviews were supplemented by a wide range of externally commissioned research and policy reviews and investigations into planning, such as the Royal Commission on Environment Pollution's report *Environmental Planning* (RCEP, 2002), the National Audit Office's report on speeding up planning applications (NAO, 2008) and the Audit Commission's reports on development control (Audit Commission, 2002) and *The Planning System* (Audit Commission, 2006b). Various parliamentary committees and bodies also produced voluminous reports and recommendations (e.g. HoC CLGC, 2008; HoC PAC, 2009).

Layered onto this distinction between developing planning and development control is the need to recognise that planning practice varies as national priorities are mediated in the light of local circumstances and priorities. While the outcomes of change over the Labour era could never be expected to be uniform, there was significant variation in impact and implementation. However, some forms of change have been more successfully implemented than others. This combination and patterning of both continuity and change is not random or accidental but related to the characteristics of planning activity, the organisational and institutional contexts within which it is located and the role of professionals. I develop an understanding of such processes and outcomes in the next chapter.

According to some, globalisation has been and is the most significant factor in shaping Labour party thinking since the early 1990s: '... the significance of globalisation and claims made about globalisation to the political economy of New Labour can scarcely be overstated' (Hay, 1999: 31). Yet it is difficult if not impossible to assess the New Labour project(s) for planning a priori. As with the New Right, there has been a difference between rhetoric and action, policy and implementation. This is especially the case in an area such as planning that is highly dependent upon the interpretation/creation of policy in a networked policy community drawing upon the professional discretion of planners. The broad theme of this chapter has been that the approach to planning evolved and was subject to a range of influences, including some broad themes or perspectives. The aim of the next chapter, before we move on to explore evidence of this in the four empirical chapters, is to provide a framework for understanding how such macro-level influences are mediated with more local and professional concerns and priorities.

3
UNDERSTANDING PLANNING UNDER LABOUR

Introduction

Chapter 2 highlighted some different views on the nature of New Labour and how planning evolved and changed during the Labour era. There is a diversity of perspectives that vary depending upon the focus (e.g. policy sector), the time period and the normative standpoint of those undertaking the assessment. This chapter develops a framework to help explain the nature of change in planning during this period. It is easy to place the practices and changes within broader literatures on, for example, governance and policy networks and to depict planning as a sub-set of such accounts. Planning is, after all, embedded within a range of wider, evolving contexts, including state restructuring (Held, 2005; Beck, 2002) and rescaling (Jessop, 2002a; Jones, 2001; Goodwin *et al.*, 2005) and shifts towards what Stoker terms the 'post-elected, local government era' (2004: 9). However, we should be wary of the lure of reductionist approaches to planning. Such contexts provide an important though partial understanding, as planning is unique in a number of important respects.

The notion of governance has been deployed as a way of understanding the changing nature of the state from one of provision to one of facilitation and coordination. Under this view of the state, elected local authorities have become just one of a number of local bodies 'governing' (Haughton *et al.*, 2010). Related elements to the notion of governance are the implications for the nature of change and transformation in policy and the role of different actors and institutions. Rather than 'top-down' government, we are now encouraged to think about networks of public, private and voluntary bodies operating at different, multiple scales, often linked through partnerships and new styles of decision-making based around negotiation and consensus-building. Policy networks, policy communities and advocacy coalitions assume a heightened significance, particularly in relational understandings that focus upon the complex networks in which policy and action are linked in formal and informal ways (Healey, 2007). Such 'institutional assemblages', as Allen and Cochrane (2007) term them, govern and manage places through a range of mechanisms.

As I discussed in the previous chapter, one understanding of the role of planning under the latter part of the Labour era was to coordinate such networked bodies and provide some form of 'governance glue' or 'meta-governance' (Allmendinger and Haughton, 2009a). However, it is clear that the notion of networked governance does not apply equally to all sectors (Haughton *et al.*, 2010). In planning there is a clear tension between the observation of fragmenting, networked forms of governance and the hierarchical, 'command and control' practices that conform to more traditional ideas of government (Cowell and Murdoch, 1999; Counsell and Haughton, 2003). For example, despite an emphasis upon 'localism', there remains an influential 'policy cascade' from the centre to the local that seeks to ensure consistency. This is backed up by reserve powers that enable the secretary of state to ensure that a particular locality does not stray too far from the policies of the government of the day and delivers upon centrally directed policy objectives and targets (Haughton *et al.*, 2010). While there is scope for autonomous local politics to interpret and determine, for example, a particular national objective in the light of local circumstances, the need to ensure that the land-use implications of nationally important issues such as climate change are not overlooked or ignored at the local level is critical. Such tensions between national and local interests characterise planning in the UK. The introduction of the Infrastructure Planning Commission was an attempt to rebalance the national need for major infrastructure, including energy, with widespread local opposition to wind-farms or new ports. The Conservatives' Planning Green Paper of 2010 directly addressed this tension by claiming that:

> The Government's approach of retaining strong central control over planning means that, in many cases, people feel that they have no say over development taking place in their areas. Local communities feel that their views are being ignored and that they are having development imposed upon them. All too predictably this sense of disenfranchisement often leads to antagonism. The result is an inherently adversarial system with opposing parties spending large amounts of time and money fighting each other, rather than seeking an agreed solution.
>
> *(Conservative Party, 2010: 4)*

The point in relation to debates around the notion of governance is that we should be wary of claims that power has been ceded from the centre and is now to be found in a miasma of multi-scalar networks. The centre, in planning at least, still dominates. This is not to deny the moves towards and understandings of governance but to temper it particularly because, as Bevir (2005) has pointed out, some use the notion of governance in a normative as well as an analytical sense, wanting a shift towards what they see as a desirable approach that devolves power and influence. The boundary between the normative and the analytical in social sciences is a blurred one, though some in planning practice find the notion of governance as a description of and aspiration for practice unhelpful at best and misleading at worst.

The portrayal of planning as a form of spatial governance is not helped by the tendency to conflate different functions and outputs. As I pointed out in the last chapter, planning in the UK separates 'plan-making' from 'decision-taking', notwithstanding the seemingly strong statutory link between the two and the desire of government to bridge the gap (DCLG, 2010b). There is a tendency, particularly among academics, to conflate these two separate functions into a homogeneous and uncontroversial unity while, in reality, the gap between the two can be significant in a number of ways. Under Labour these two 'arms' of planning (which is itself a simplification) were approached (or 'modernised', to use New Labour terminology) in different ways. The 2004 Planning and Compulsory Purchase Act focused mainly upon development planning, and subsequent discourses around 'spatial' planning emphasised governance notions of coordinated networks of actors involved in 'place shaping and making'. Development control, on the other hand, was subject to performance indicators and targets. The assumption was that there was a seamless connection between the two and that the 'plan-led' approach would ensure that development control would be residualised into a largely administrative function (Allmendinger, 2006). The reality was that development control became highly politicised as the implications of growth in general, and development schemes in particular, became apparent to local communities and stakeholders. As challenges to proposals and decisions became more common, attention was drawn to the semi-judicial nature of this element of planning. The analyses and proposals in the Planning White Paper (DCLG, 2007a) and the provisions of the 2008 Planning Act recognised the naivety of the idea that decisions on development would simply flow uninterrupted from development plans. While plan-making has better embraced the notions of governance and fragmentation, development control remains much more embedded within the notions of regulation and government.

The further defining characteristic of planning that needs to underpin any framework for understanding is the requirement to engage with the difference between the formal and the informal: 'The "power" of planning lies not in its formal procedures, its legislative foundations or its political role, but in the communicative practices of the social relations in which planning is entangled' (Hillier and Healey, 2008: 3). There is a rich body of work that explores the informal and formal elements of planning in relation to language and communication and the 'framing' of decisions (e.g. Schön and Rein, 1994; Hajer, 1989; Healey, 1996; Gunder and Hillier, 2009). A related school of work has highlighted the more nefarious abuses of language and power through focusing upon rationality (e.g. Flyvbjerg, 1998), while others have explored how planning hegemonies and related doctrines are used to achieve compliance with agendas that favour particular interests (e.g. Coop and Thomas, 2007).

Thus, it is important to recognise the discourses employed by Labour around planning, the informal, communicative practices of planning, and the relationship between the two. Again, this is not necessarily a unique characteristic of the Labour era. However, there is an emerging body of thought that highlights how 'Third Way' politics and the emphasis upon grand discourses such as 'sustainable development', 'smart growth' and 'community', allied with the use of partnerships and other consensus-building techniques, can serve a neoliberal growth agenda while excluding those who offer alternative visions (Raco, 2005). Labour attempted to create a hegemony around growth and used discourses such as 'urban renaissance' to present development as inevitable and unquestionable. Planning was reorientated from a conflict mediation role to focus upon how best to achieve growth and in what ways – for example, 'sustainably'. Under this view, planning, particularly in the latter part of the Labour era, can be seen as a managerial or administrative process that seeks to smooth over dissent, critique and conflict through all-embracing, 'fuzzy concepts' such as 'sustainable development' (Haughton and Allmendinger, forthcoming, 2010).

Any understanding of this era needs to engage with the realities and uniqueness of planning under Labour as it actually operated, not as some wished it had or as it might appear in locating it within normative frameworks of understanding. This is not to say that realities of planning under Labour were unique to that era. What was unique were the purposes to which planning was put, the role that planning and planners played, the acquiescence of those involved, and the mismatch between the rhetoric of 'spatial planning' and 'sustainable development' and the reality of a growing anti-development attitude, particularly in large swathes of southern England. Less distinctive were the differences between intended and unintended outcomes, debates over the appropriate scale for planning, and how to balance competing demands around certainty and flexibility, coordination and speed (Haughton *et al.*, 2010). These issues and the characteristics of planning discussed above provide a backdrop to how we make sense of change, but they do not, in themselves, provide a theory or framework of change. Instead, they point us in the direction of how best to conceptualise such issues.

One of the criticisms that could be aimed at recent evaluations of planning and spatial planning is that they have largely assumed an automatic transfer of policy and rhetoric into practice and have too easily taken at face value the claims of change. A further criticism, and one that will emerge through the subsequent chapters in this book, is that accounts of change thus far have mostly concentrated on rhetoric rather than looking at outputs in practice. Not only does this present a misleading picture, it also necessitates an understanding that accounts for how ideas and policies emerge and are translated and mediated and what factors influence interpretations. In addition, it requires assessment of the full panoply of intended and unintended consequences and outcomes. Those who have looked at planning practice and sought to explain the gap between expectation and reality have themselves pointed to the need to understand better the 'black box' of local practices and politics (e.g. Newman, 2008). Consequently, if we are to understand change in planning as being related to outcome rather than rhetoric, then the hub of concern needs to be on the institutions of planning and those who work within them, particularly at the local level.

As discussed above, any framework for understanding should not be imposed or 'bought in' from other sectors or studies but be adapted to account for the unique characteristics of planning. Of particular importance is the need to account for the institutional context of planning and the emergent and evolving practices that shape outcomes.

Understanding change in planning and governance

This study of planning during the New Labour years is framed by two related areas of understanding. First, I draw upon New Institutionalism as a way of comprehending transformations in policy and ideas within a fragmented landscape of formal and informal networks of individuals and groups. Second, within New Institutionalism I emphasise the role and significance of discourse. Various discourses around growth in the south of England (e.g. the sustainable communities programme) and within cities (e.g. the 'urban renaissance' agenda) were woven into 'win–win–win' outcomes around environmental, economic and social benefits. Labour's deployment of discourses to create a consensus-based growth strategy echoes the concerns of those who highlight the changing role of the state in managing dissent. During the Labour years, planning was refocused from an arena that mediated and sought to reconcile different objectives and views to one where dissenting voices were marginalised as an aberration from mainstream discourses. Discourses such as 'sustainable development' achieved widespread 'buy-in' from groups and individuals by conjuring up images of uncontentious growth. But it was growth all the same. The experiences of planning under New Labour place a heightened significance on discourses within a New Institutionalist framework.

New Institutionalism

Institutionalism has become a common framework within which to explore and better understand local governance (e.g. Stoker, 2004; Lowndes, 2005), planning (e.g. Vigar *et al.*, 2000; Healey, 2006a; 2007) and development and property markets (e.g. Healey, 1992; 1994; Guy and Henneberry, 2000; 2002; Ball, 1998). It is not a single body of theory, and the emphases and application of institutionalism will depend upon the focus of any study. Planning in the UK has a number of characteristics that point towards the need to understand change as being less about government rhetoric and desire and more about relational mediation, interpretation and creativity in policy. New Institutionalism puts the actor (planner) and local circumstances and contingencies back into the frame of understanding. As an activity planning is particularly suited for such an understanding, given its key characteristics.

- *Multiple, overlapping policy processes* Planning in the UK combines different policy processes that are highly political in terms of their distributional consequences. Planning is composed of a number of overlapping and connected processes, including semi-judicial processes (e.g. preparation and adoption of development plans, local inquiries, etc.), technical, expert-driven processes (e.g. retail impact assessments) and engaged, consultative processes concerned with public and other involvement. The relationships among these processes, when and where they are used, and the significance placed upon them vary within and between localities. Who chooses such processes and in what circumstances they are used is not simply a reflection of formal responsibilities and roles but a negotiated outcome from a range of factors.
- *The role of doctrine and flexibility* Planning within local planning authorities and private practice is undertaken by a cadre of professionals. Planners employ certain doctrines concerning spatial arrangements, future developments and the longer-term legitimacy of planning action and scope (Hajer, 1989, 1995; Alexander and Faludi, 1996; Needham, 1996; Murdoch and Abram, 2002). At one level doctrines can act like discourses in that they serve powerful elites, and planning can be used to further the interests of the few. But a subtler notion of planning doctrine highlights the necessity for planners themselves to 'carve out' a malleable agreement among interests to facilitate potentially radical change in familiar terms (Coop and Thomas, 2007). Attention needs to be focused upon challenges to such doctrines both from within (e.g. new professional ideas and practices) and outwith (e.g. national policy changes, new institutional structures, etc.) planning. Experience from the 1980s highlights how the profession and local authorities were able to accommodate, deflect and even 'turn' such challenges to existing practices and ideas (Allmendinger and Thomas, 1998). However, the 'renaissance' of planning during the Labour years was accompanied by new, more positive roles and ideas for planning.
- *Multiple and competing objectives and rationales* While professional discourses and doctrines help shape the practices and scope of planning at different scales, these compete within a context of changing national political discourses. Planning has been subject to changing rationales over time (see Haughton *et al.*, 2010: ch. 2) and its purpose and objectives have varied and include minimising negative externalities, the provision of public goods, the redistribution of wealth, and the creation of environments in the public interest. As Thornley (1991) points out, these varying rationales have existed simultaneously and have been used selectively and often by the same individuals to justify a particular course of action (see also Reade, 1987). What marks out the recent period of planning as distinct are the multiple, competing and seemingly irreconcilable *national* discourses of planning. I discuss these in more detail further on in this chapter as being of particular significance in understanding change. However, the point here is that these competing discourses reflect deep-seated struggles over the purpose of planning within government.

- *The significance of professionals and discretion*   Discretion is a traditional characteristic and a necessity of local government. Authorities can act 'as they think fit' in the 'public interest' providing their actions are not *ultra vires* unreasonable or procedurally improper. While this flexibility is circumscribed, professional discretion adds a further dimension. Legislation has been drafted in such a way as to allow considerable discretion in what is a 'material consideration' in planning (Booth, 2007). Discretion and the notion of planners as professionals with insight and expertise in the subject combine to carve out a powerful potential to interpret and thereby create policy either in conjunction or at odds with local political and public desire. This adds further significance to the role of doctrines and also emphasises the need to understand the role of norms and networks within local governance and planning practice.
- *Embeddedness within local governance at multiple scales*   It is now common to refer to multi-scalar planning and governance. However, planning in the UK has, to a greater or lesser degree, always operated across and between scales in a networked fashion within public bodies (i.e. across professionally aligned departmental boundaries) and between a range of public, private and other agencies. The multiple and shifting scales of what we term 'statutory' or 'formal' planning (e.g. Regional Spatial Strategies, Sustainable Community Strategies, etc.) exist alongside and interact in a variety of ways with 'informal' or 'soft' spaces of planning (Allmendinger and Haughton, 2009b). These soft spaces become temporary and functional approaches that often seek to 'short-circuit' and address what are perceived as time-consuming and cumbersome statutory plans and strategies that can be and often are focused upon scales and administrative boundaries that do not map onto functional planning areas or development opportunities.

The above characteristics highlight that planning practices are multi-faceted and composed of complex formal and informal activities. Individuals, therefore, have a key role within practice networking planning through scales and within organisations in highly political, locally distinct contexts.

Where institutionalism and New Institutionalism differ is in the emphasis placed upon the significance of informal rules. New Institutional analyses within local government highlight the struggles to introduce new policies and processes and emphasise the need to understand change and continuity less through formal structures and processes and more through the actual 'rules of the game', which encompass 'values, norms, interests, identities and beliefs' (March and Olsen, 1984: 17) and shape relations, interpretations and outcomes in public administration. The experiences of planning under Labour exhibited the significance of such a focus in helping understand change.

A range of New Institutionalisms exist that reflect different emphases and understandings of change and actor behaviour. Two particular approaches stand out. Lowndes (2005) has developed an approach which conceptualises local governance as an 'institutional matrix' made up of distinct but interacting rule sets. Such rule sets are the 'rules of the game', either consciously designed (e.g. Local Development Frameworks) or informal (e.g. unwritten customs or codes). Informal rules can support both positive patterns of behaviour (e.g. public service ethos) and negative patterns (e.g. departmentalism). Rules are employed by different 'players' in the game, whether organisations (e.g. local authorities, political parties, voluntary bodies and private companies, etc.) or individuals (planners, politicians, business people, community activists, etc.) (ibid.: 293).

Another flavour to New Institutionalism is provided by Schmidt (2008), who has concentrated on the role of discourses within an institutionalist framework. For Schmidt, discourses include not only what is said but to whom and in what context. She focuses attention within institutionalist analyses upon the ideational background of actors and how these are conveyed into daily practices of political action. The difference between the two approaches is really one of emphasis. However, both provide potentially useful frameworks for understanding planning and change under Labour. Lowndes's strategy helps explain why change is difficult, while Schmidt's more dynamic model explains and even assumes change.

Central to both approaches is a distinction between what Lowndes identifies as her four types of rule set and Schmidt terms her three types of idea. Lowndes argues that the rule sets of local governance are comprised of different spheres – the political, managerial, professional and constitutional – and are 'nested or embedded within wider institutional frameworks that exist above, below and alongside local government itself' (2005: 294). Rule sets evolve and change at different speeds and can either reinforce or contradict each other under a range of influences, from national government, the media, and locally specific conventions and cultures. In planning, local rule sets could include strong professional autonomy within a hung council combined with an anti-development stance within the wider community. This can lead to an effective 'planning style', which I discuss further on in this chapter. Rule sets can become reinforced, for example, through a particular interpretation of housing targets and national brownfield development targets in an emerging development plan. Given the professional autonomy involved, local political or managerial rule sets which could be pro-development would struggle to effect change. Despite attempts to increase housing delivery at a regional or national level, this particular alignment of rule sets could lead to a different outcome.

As Lowndes emphasises, however, rule sets do not operate only at the local level. Going back to the example above, if formal government attempts to increase housing supply through targets at a national or regional level fail, then other, more informal rules can achieve these objectives. The need to ensure a five-year supply of housing land in any area provides a way of ensuring that local resistance (or 'creative interpretation') does not thwart national priorities. If a five-year supply of land is not available, then local planning authorities and, more significantly, the Planning Inspectorate should look 'favourably' on proposals. This introduces a 'presumption in favour of development' within the wider and statutorily based 'presumption in favour of the development plan'. In other words, rule sets are evolving to overcome other rule sets, which will, in turn, evolve themselves.

Lowndes's approach privileges and helps explain inertia or continuity in policy. Schmidt, on the other hand, develops a New Institutionalist framework that, she claims, helps explain change. In her conception, ideas exist at three levels of generality. At the first level are *policy solutions* proposed by policy-makers. The second level concerns *paradigms*, which reflect the underlying assumptions or organising principles orientating policy. Paradigms help define the problems to be solved by policies, the issues to be considered, the goals and the methods or instruments to be used. Finally, *public philosophies* present deep world-views, or *Weltanschauungen*, that underpin both policies and paradigms through providing values and principles of knowledge and society. While policies and paradigms are regularly debated, public philosophies are far more immutable.

Distinguishing between different levels of ideas is a common approach in understanding change in public bodies (see Giddens, 1984; Healey, 2004a, 2007; Schön and Rein, 1994). However, Schmidt goes on to distinguish between two types of ideas within policies, paradigms and philosophies — cognitive and normative. The former concerns 'what is and what to do' while the latter relate to 'what is good or bad' about what we should do. Cognitive ideas help provide solutions for policies, while paradigms shape the problems to be solved and help identify methods by which to solve them. Normative ideas help provide the justification for cognitive ideas at policy and paradigmatic levels by attaching values to them.

In planning, day-to-day attention is focused on policies and their creation/interpretation. We can also identify policies that are more ephemeral (e.g. the emphasis on increasing housing density) and others that are akin to paradigms (e.g. green belts). Other distinctions that map onto Schmidt's approach are the significance of national policy and the objectives of planning (e.g. the creation of sustainable communities) that constitute paradigms. Both fit within relatively enduring and influential public philosophies of planning. During the 1980s the public philosophy of planning was pro-development, market shaping, then during the 1990s it became much more concerned with climate change. We can also see how the notion spatial planning seemingly fits into this framework.

Both Lowndes and Schmidt provide different flavours of New Institutionalism. In our evaluation of change we also need to understand better how change occurs and how variability in change can be explained under these two New Institutionalist frameworks. Under Schmidt's approach current understandings do not successfully account for how new policies emerge and replace existing ones, nor do they properly explain how 'bad' policies succeed in being used and 'better' ones do not. Rather, discourse is taken to be an institutionalised structure of meaning that channels political thought and action (Schmidt, 2008: 309). Discourses concern not only what is said but to whom, how and where. They can operate at different levels of ideas (policy, paradigm, public philosophy) and different types of ideas (cognitive and normative). There are two distinct areas of focus in relation to discourse and public administration. The first is the policy sphere, where coordination about policy discourses between actors and groups over the creation, elaboration and policy and programmatic ideas is undertaken. This may be done through, for example, epistemic communities, advocacy coalitions, discourse coalitions or advocacy networks. Such coordination may involve 'policy entrepreneurs' or 'mediators' who become catalysts for change. In the political sphere, communicative discourse involves the presentation and legitimation of ideas to the general public. Communicative discourse is undertaken not only by those in power but also by opposition parties, experts, think tanks, etc. Nor is communicative discourse a one-way process, as there is feedback and evolution through engagement and public involvement.

The ability of discourses to effect change relies on their being relevant to the issues at hand, adequate, applicable, appropriate and having resonance (Schmidt, 2008: 311). However, to be persuasive they should also be consistent and coherent across policy sectors. This does not mean that they have to be explicit, as a degree of vagueness is often useful in developing appeal. Discursive interaction and processes usually involve both arguing and bargaining and can be vehicles for elite domination and power. Two final dimensions are worth adding. First, Schmidt's approach allows for agents to be both within and outwith their institutional contexts. Agents have 'background' ideational abilities. By this she means that agents can 'fit in', be part of existing institutional contexts and contribute to them. At the same time they also have 'foreground' discursive abilities through which they may change their institutions. In other words, through the combination of background and foreground, agents are able to 'think, speak and act outside their institutions even as they are inside them, to deliberate about institutional rules even as they use them and to persuade one another to change those institutions or maintain them' (ibid.: 314). Professional codes and planning doctrines as well as local cultures and preferences provide powerful institutional 'background' discourses and influence planning policies and paradigms. Yet the implementation and roll-out of such policies and paradigms is not uniform, and there are debates over how local circumstances interact with national paradigms and policies.

Lowndes's understanding of change from her New Institutional perspective provides a different flavour. For her, power relations help shape the development of institutions and institutional change is never a technical matter. In fact, purposive attempts at institutional change are hard to achieve, as they usually involve challenging such power relations (2005: 294). She follows Ostrom's (1999) distinction between 'rules in use' and 'rules in form', which recognises that apparent institutional change may be 'skin deep', with older rule sets continuing to dominate. Drawing upon the work of, among others, Peter John (1999), Lowndes develops a biological analogy and posits an evolutionary understanding of institutional innovation. A key element in institutional innovation is the meme. Memes are a central, directing idea or policy within an institution that, like genes, transmit ideas, though they do so in ways that allow for evolutionary change (adaptation) and sudden change (punctuated evolution). Memes are selected from a 'pool' of ideas by ideological deliberation and 'carried' by actors and interest groups who advocate in their favour and promote them. They continue to evolve until they no longer adequately explain or help frame a situation. The point at which memes can no longer adapt or evolve provides an opportunity for 'institutional entrepreneurs' (Lowndes, 2005: 297) to advance new explanations.

Such a situation arose towards the end of the 1970s and the long postwar consensus when there was a fundamental (punctuated) shift in ideas around the role of the state by the New Right and its associated think tanks. These ideas were not in themselves new. Instead, according to this view, the 'primeval soup' of ideas provided an opportunity for an existing frame that took advantage of the situation and emerged. The mechanism of 'choosing' is less well articulated, though two aspects seem crucial. First, the role of persuasion and, ultimately, the securing of a parliamentary majority is a key prerequisite to punctuated evolution. The other is more pragmatic, as related initiatives and approaches are 'tested' and found successful or not. As an example, Thatcherism could logically have a range of approaches to local government management and, through an approach of trial and error, a range of initiatives were adopted that amounted to what has been termed 'new public management'.

Such punctuated change is the exception. 'Normal evolution' involves actors making and remaking institutions on a daily basis. In day-to-day practices there is room for creative interpretation between institutional stability and volatility. There are always areas of ambiguity in the interpretation, allowing the 'bending' of rules to meet the circumstances of a particular situation. Such small step-changes can amount to larger changes over time. The focus of attention in daily practices is on the role of entrepreneurs and 'institutional emergence' (Lowndes, 2005: 299) that can involve the recombination and reshuffling of existing components or institutional material. Lowndes identifies three strategies through which institutional entrepreneurship occurs.

- *Institutional remembering*   The emergence of new ideas, institutional rules and tools does not eradicate older resources but places them back into the pool of potential approaches. The significance of institutional remembering is the way in which it can be used to provide gravitas or a convenient back-story to a particular initiative when an institution is reactivated. Imbuing a new institution with such a legacy can be persuasive in securing acceptance among sceptics. The rehabilitation of elected mayors and town and parish councils is advanced as an example of institutional remembering under Labour.
- *Institutional borrowing*   This occurs when actors are involved in more than one rule set – for example, managerial and political – and involves the transfer of ideas from one rule set to another, particularly in situations where one rule set evolves at a slower pace than another. In the case of management rules evolving more quickly than political responses, the institutional 'stickiness' of the latter can be tackled by institutional entrepreneurs through applying the rules of the former. Thus, local councillors have been subject to training and mentoring to combine managerial and political leadership.
- *Institutional sharing*   This is a more outward-looking approach that involves expanding the repertoire of resources and approaches from other sectors and

actors, which can be facilitated by the more porous and overlapping relations between networks within which actors are involved. A further impetus to sharing can be the increasing use and influence of information technology, which has allowed experiences to be shared across sectors more easily.

Both Lowndes and Schmidt posit an active, significant and potentially darker role for agents. In acknowledging the persuasive and entrepreneurial dimensions, allowing for cognitive dissonance and the appearance of change as opposed to actual change, they come close to identifying the multiple roles and discourses that shape the day-to-day activities of planning and planners. Planners work within structures but change those structures through action. They can, and do, argue and bargain, persuade and manipulate in ways that lead to both change and continuity. In positing both background and foreground ideational abilities, Schmidt specifically allows for and even expects strategic behaviour and complex, multiple discourses operating both consciously and unconsciously.

Discourses and cultures of planning

Within planning there has been empirical work that explores the daily practices of planners and others within an institutional framework (see, for example, Healey, 1998, 2006a, 2007; Vigar *et al.*, 2000). However, in order to understand better the dynamics of planning under Labour, three areas require some further attention: first, the role, significance and understanding of discourses within planning during this era; second, the notion of planning cultures; and, finally, the scope for and operation of professional and administrative discretion.

As discussed above, a focus upon discourse is concerned with how ideas originate, evolve and are used to shape action. Discourses are taken to have a causal influence on institutional transformation through providing frames of reference that help policy-makers to identify issues and propose possible solutions (Schön and Rein, 1994). In an area such as planning, discourses have long been used as a framework through which to understand the multiple and sometimes competing objectives and the origin of particular policies and practices/methods (e.g. Underwood, 1980; Brindley *et al.*, 1996). Such policy discourses are developed and articulated by policy communities within policy networks through a variety of means (Vigar *et al.*, 2000). However, while there is a growing awareness of the significance of discourses within planning, the relationship between the different levels of discourse, discussed by Schmidt above, becomes more critical because of the largely centralised nature of UK planning. Different policy areas will need to emphasise particular elements of the New Institutionalist framework for them to 'make sense'.

One characteristic of the latter half of the Labour era was the lack of progress in putting development plans in place. By late 2009, only 20 per cent of local authorities had adopted their Local Development Framework Core Strategies, five years after the 2004 Act introduced the new system of development planning. Given the 'plan-led' approach that has underpinned planning since 1991, where individual decisions have to be made in accordance with the development plan unless there are material reasons not to do so, one obvious question is: In a period of high development activity and without widespread plan coverage, what criteria have individual applications been assessed against? The answer will, of course, depend upon the nature of individual proposals and localities. However, at an aggregate level the answer was a combination of aging local policy frameworks, emergent thinking, local planning doctrine and, most significantly, national planning policy.

There are strong arguments that the introduction of the plan-led approach was an attempt to centralise local planning policy in any case (Allmendinger and Tewdwr-Jones, 2000; see also Chapter 1). However, from the end of 2004 until the latter part of 2006 there was an eruption in central planning guidance and information. This was in addition to the wide range of bodies and initiatives that were introduced around the new system of development planning and performance management in local government. The upshot, as the Barker review (Barker, 2006a) noted, was complexity, confusion and delay – all points that the government later accepted (DCLG, 2007a). As Barker further noted, despite attempts to reverse the growth in central government planning guidance since (at least) 2002, there were still over 830 pages of national planning policy in 2006 (Barker, 2006b: 98). Part of the reason for this growth was the enlargement in the range of issues that planning had been charged with addressing. As Table 3.1 highlights, planning became embroiled in an increasing range of issues and objectives.

A second source of growth was the need to explain the new system of development plans, which involved the publication of companion guides as well as various retrospective attempts to plug gaps in knowledge and process that were previously thought to be understood (DCLG, 2010b).

Finally, there was the overarching notion of the shift in the purpose and processes of planning that was labelled 'spatial planning'. I discuss this more in Chapter 5. Suffice to say here that spatial planning amounted to a new discourse that highlighted certain aspects of New Labour's spatial governance, namely, sectoral coordination of multiple issues around places at different scales. Again, much of the national planning policy that emerged after the 2004 Planning and Compulsory Purchase Act emphasised the shift to this approach, though in ways that provided broad, overarching objectives, aspirations and themes. Planners in the public and private sectors were therefore simultaneously trying to grapple with a new system of development plans, a new notion of planning, and significant expansion in central planning guidance against a backdrop of heightened development activity and staff shortages.

In this environment of uncertainty it was possible to find an interpretation of guidance or advice to support arguments for a range of positions, and local authorities, developers and others struggled to uncover the overall purpose of planning, particularly as far as development plans were concerned. Around these multitudinous changes and positions were the 'frames' of discourses that, according to the mainstream view, helped give New Labour's approach shape and direction and

TABLE 3.1 Areas where local planning authorities have been given additional or more complex responsibilities since the Town and Country Planning Act 1990

Issue	Sources
Access for disabled people	PPG1: 1992; Circular 11/95; PPG1: 1997
Affordable housing	PPG3: 1992; Circular 11/95
Air quality	PPG23: 1994; Circular 15/97
Archaeological protection	PPG16: 1990
Contaminated land	PPG23: 1994; Circular 02/00; PPS23: 2004
Crime prevention	Circular 5/94
Design of buildings	PPG1: 1997
Environmental Impact Assessment	Regulations 1998
Gambling	Gambling Act 2005
Gypsy and traveller sites	Circular 18/94 (update)
Housing in multiple occupation	Circular 12/93
Licensing	PPG6: 1996; Licensing Act 2003
Nature conservation	PPG9: 1994; PPS9: 2005
Noise	PPG24: 1994; Circular 11/95
Planning obligations	Circular 1/97
Pollution controls	Environment Protection Act 1990; PPG23: 1994
Retail	PPS6: 1996
Sustainable development	PPG1: 1992; PPS1: 2005
Telecommunications	PPG8: 1992
Transport	PPG13: 2001
Waste	PPG23: 1994; PPG10: 1999; PPS10: 2006

Source: Barker (2006b: 69)

made sense of the complexity. Such discourses assumed heightened importance given the ambiguous policy framework. Two issues arise: first, what were the discourses under Labour and, second, what were the broader, structuring forces that helped shape them?

There were a number of simultaneous, competing and potentially irreconcilable discourses that impinged upon and directed planning during the Labour years. These are different to though have some connections with the competing views of planning itself outlined in Chapter 2.

- *A key mechanism in tackling climate change* The UK Strategy on Climate Change (DEFRA, 2005) identified the key role spatial planning could make in tackling this problem – a role that drove spatial planning policy at national, regional and local levels (ODPM, 2005a). This discourse posits a strong, regulatory role for spatial planning, highlighting the need to assess stringently the locations of new development against sustainable development criteria and limitations upon 'inappropriate development' – for example, in areas liable to flooding. It privileges environmental over social and economic concerns in the triad of sustainable development.

- *An inclusive, integrative process*   Key to the process of spatial planning is the notion of involvement and 'buy-in' from a range of stakeholders. Statements of Community Involvement helped emphasise the notion of partnership (Morphet, 2008) while sectoral policy integration was partly achieved through mechanisms such as Local Strategic Partnerships (ODPM, 2005b). The accent in this discourse was upon the need to ensure that processes of spatial planning were part of a networked form of governance providing a spatial expression to wider community and sectoral aspirations.
- *A brake on economic development and competiveness*   The UK Treasury led the charge that planning regulations and controls inhibited development and competitiveness. Initial concerns over impacts upon consumer prices and competition in the hotel and supermarket sectors (McKinsey, 1998) were followed by macro-economic concerns over housing supply, particularly in the south-east (Barker, 2004), and then wider anxiety over economic globalisation and competitiveness (Barker, 2006b). More specific and less fundamental concerns over regulatory impacts also came in the form of an inquiry into transaction costs (Killian and Pretty, 2008) and the need for a more positive 'culture change' in planning (DCLG, 2010b). This discourse was not new, given its prominence during the more neoliberal attitude towards planning during the 1980s. However, it served to privilege economic development and commercial activities and helped justify growth and development.
- *Addressing reactive, wider public concerns over a range of policy issues*   In the normal course of events a range of issues emerge which require politicians and the government to react and 'do something'. Among examples during the New Labour era were obesity (HoC HC, 2004), terrorism (Home Office, 2009) and children's play (Sport England, 2009), all of which added to the extensive list of issues that were regarded as being legitimate planning policy objectives (see Table 3.1). There was an expectation that planning could address these either through regulation or through some form of fee or tax to mitigate any impact. This discourse of planning positioned land and property development as a nexus for addressing a range of social issues, particularly through the submission of 'expert' evidence to justify and support proposals to risk-averse local authorities.
- *A tool of the 'urban renaissance'*   The push for regeneration of towns and cities through an approach that combined focused neighbourhood renewal and social inclusion provided planning with a crucial role. In a positive guise, planning offers the spatial expression and process for joining up multiple sectors and objectives as well as being the focus for community engagement. Notions around 'mixed communities', the 'public realm' and a 'sense of identity' underpinned an unsubtle reshaping of urban areas to attract the middle classes back to city-centre living (Amin et al., 2000; Lees, 2003; Atkinson, 2004). Such a 'cappuccino urban policy' contrasted with the 'negative' role for planning in both restricting urban sprawl and helping 'plan out crime' and 'disruptive elements' such as the homeless, beggars and teenagers (MacLeod, 2002).

While discourses are generally posited as 'frameworks of understanding' from which actions are understood and 'flow', this is at odds with the more political and persuasive understandings of the role of ideas in shaping arguments and actions within dynamic and contested institutional networks and contexts. This latter perspective is more akin to the situation under Labour, where policies and planning discourses were far from clear-cut. Planning discourses under Labour were related more to the notion of hegemony, particularly around the promotional and persuasive nature that sought to make real its inclusive 'big tent' approach. As such there are some differences with how such discourses were employed. First, concepts such as 'sustainable development', 'smart growth' and 'urban renaissance' were deployed in a deliberate attempt to create an environment of acquiescence. They allowed different, seemingly opposing policy communities to sign up to a notion, such as sustainable development, that eluded and suppressed conflict and opposition. A key characteristic of Labour's planning discourses was their malleability to different groups and circumstances. Thus, the Home Builders Federation, the Confederation of British Industry, the Campaign for the Protection of Rural England and Friends of the Earth could all sign up to the notion of sustainable development, each thinking that it meant something subtly different. Where the balance or emphasis actually lay could, however, be significant. An early rationale for the second Barker inquiry into planning concerned the balance between environmental, economic and social dimensions of sustainable development, the clear implication being a need to redress the dominance of environmental interpretations with a renewed economic emphasis.

The second major difference was that Labour, with support from the planning profession and academia, reorientated the processes and purpose of planning to manage and administer such discourses. From around 2003/4 onwards it became clear that an opportunity had opened up in within central government's consideration of the future of planning. From a largely suspicious and even hostile attitude, a loose coalition of interests began to portray planning as a nexus of New Labour's disparate policy concerns around, *inter alia*, housing affordability, climate change and economic competitiveness. A more supportive secretary of state, John Prescott, was willing to accept the argument advanced by the profession and others that planning could act as a form of 'meta-governance' through the notion of 'spatial planning'. Thus, planning became deeply implicated in the creation and deployment of Labour's discourses.

These two differences were significant, though a more subtle difference between Labour's multiple discourses was one of intention. Rather than being portrayed as an arena in which competing and often irreconcilable objectives could be addressed in an open and democratic way, planning under New Labour eschewed politics and sought instead to present a range of discourses that captured competing views and with which it was difficult to object (who could not agree to the notions of 'sustainable' development, or 'smart' growth? The idea that someone was for 'dumb' growth naturally didn't arise). Planning, or, more accurately, spatial planning, became concerned with 'growth management' which, when combined with tools or approaches such as 'partnership', began to undermine the public interest notions of planning.

Again, the notion of hegemony is not new to planning, but the combination of hegemonic discourses with a reorientated and refocused planning into a mechanism for avoiding political choices and debates marks a significant departure from the postwar approach. Such an understanding of (spatial) governance being concerned with managing growth through consensus echoes the emerging notions of the post-political condition (Swyngedouw, 2007). The broad thrust of post-political analyses is that the collapse of communism heralded an 'end to ideology'. Rather than making choices around socialism or liberalism, society and government needed to address questions that dealt solely with rational negotiation and decision-making based upon economic necessity (Žižek, 2000: 323). While views vary, neoliberalism can be taken to be the:

> theory of political and economic practices that proposes that human well-being can best be advanced by liberating individual and entrepreneurial freedoms and skills within a framework characterized by strong property rights, free markets, and free trade. The role of the state is to create and preserve an institutional framework appropriate to such practices.
>
> *(Harvey, 2005: 2)*

The post-political condition is founded upon the acceptance by successive governments of neoliberalism as an economic system (Mouffe, 2005; Žižek, 2000). Under such a paradigm the role of government (and, by implication, planning) is to facilitate neoliberalism. This has significant implications for the state in general and for planning in particular, changing what is understood to be within the remit of planning, who engages with the system and under what terms. Conflict and opposition remain but are assumed to exist within limits: 'Postpolitical parliamentary rule ... permits the politicization of everything and anything, but only in a non-committal way and as non-conflict' (Swyngedouw, 2007: 25). Thus, housing growth, economic competitiveness, the impacts of climate change, etc., all form discourses that planning will then seek to manage or administer. The aim is to create consensus around managing options for growth through the deployment of discourses that externalise or banish conflict. This casts a new light on the role and significance of mechanisms such as partnerships that underpinned much of New Labour's approach to governance (Geddes, 2006).

What we need to bear in mind is that planning was both an *object* of post-politics through the wider processes of local government modernisation, including performance targets, etc., and a *conduit* through which growth and other discourses were to be managed. However, one element, largely overlooked in post-political theory, is that of resistance. In planning such resistance is derived from discretion and the scope for autonomy within central–local tensions. Resistance has been a characteristic of planning under New Labour, as subsequent chapters highlight. I turn to this again in the conclusions. However, two further elements of discretion and autonomy need emphasising to understand fully the impact of changes to planning under Labour.

Culture

A useful starting point in understanding paradigmatic shifts is the idea that ideas/policies (or memes, as Lowndes terms them) can be advanced by institutional entrepreneurs who take advantage of the drawbacks or lack of adaptability of existing frames. Nevertheless, it leaves open questions over the relationship between different rule sets. Within the broad, centrally directed UK planning system, national planning discourses assume an automatic primacy. However, as the Labour era was replete with multiple and often competing discourses, the question then arises: Who decides and in what circumstances? The scope for autonomous local politics has long been a focus of study and debate, with recent understandings emphasising more networked forms of governance over more traditional, central–local relation perspectives (Rhodes, 1997; Marsh and Rhodes, 1992). It has also long been understood that the professional nature of planning has carved out discretion in how issues/ideas are interpreted. I cover discretion in more detail in the next section, but here it is worth highlighting the role and significance of planning doctrine, or what has recently been relabelled 'planning culture'.

The notion of planning doctrine is a mainstay of planning studies (e.g. Alexander and Faludi, 1996; Needham, 1996; Murdoch and Abram, 2002). However, there have been recent attempts to embed planning doctrines within more overarching local and national cultures and to move from the analytical basis of doctrine to argue that planning practice should be more locally sensitive and aligned to particular cultures of places (Sanyal, 2005). Planning culture refers to the 'different planning systems and traditions, institutional arrangements of spatial development and the broader cultural context of spatial planning and development' (Knieling and Othengrafen, 2009a: xxiv). Such cultural contexts provide a filter through which more homogenising forces related to globalisation are interpreted and acted upon and are often related to 'imagined places' that influence the practice and aims of planning in specific places (see, for example, Ward, 2004). An example of such idealised aims or places is the notion of green belts (Elson, 1986), which themselves draw upon much older and complex notions of rural health and urban blight found in the analysis of and arguments for garden cities. As relevant to us are the ways in which cultures have developed around processes and particular rationalities that privilege certain groups (planners, the well off, the well organised, etc.) over others (Healey *et al*., 1988).

The lens of doctrine or culture has been and can be used to understand the differences between planning systems and practices at a national and a regional scale. It focuses attention upon the particularities of history, attitudes, beliefs and values, cognitive frames, interpretations of planning tasks and responsibilities, political and legal traditions, rules and norms, different levels of market integration, and different institutional structures of governance at all levels of planning (Knieling and Othengrafen, 2009b: 39). A clear example of the significance of culture can be found in Brindley, Rydin and Stoker's identification of different 'planning styles' in the UK. Different localities work around specific economic and social configurations leading to distinctive political frames:

> Each style represents a particular stance in the debate on planning and proposes a particular mix of policy goals, working methods and identity for the planner. Some styles are strongly influenced by a radical vision and have the character of blueprints for local experiments. Other styles are not so new but derive from adaptations and modifications of established planning methods.
>
> *(Brindley et al., 1996: 7–8)*

Table 3.2 sets out a typology of six different planning styles during the 1980s, an era when there was supposedly a 'rolling back' of planning activity to a minimal state model (Ambrose, 1986; Thornley, 1991).

Planning styles are more than simply decision frameworks favouring one option over another, but rather involve associated institutional arrangements, rationalities and processes (Table 3.3).

In the 1990s, Brindley, Rydin and Stoker revisited their six planning styles of the 1980s and found that practice had evolved into two main styles (1996: 195), though they envisaged new styles emerging. Planning styles within different areas during the 1980s and 1990s were the foundations upon which the Labour era built. This does not undermine the centralised nature of UK planning but defines the

TABLE 3.2 A typology of planning styles

Perceived nature of urban problems	Attitude to market processes	
	Market-critical: redressing imbalances and inequalities caused by the market	Market-led: correcting inefficiencies while supporting market processes
Buoyant area: minor problems and buoyant market	Regulative planning	Trend planning
Marginal area: pockets of urban problems and potential market interest	Popular planning	Leverage planning
Derelict area: comprehensive urban problems and depressed market	Public-investment planning	Private management planning

Source: Brindley *et al.* (1996: 9).

TABLE 3.3 Characteristics of six planning styles

Planning style	Institutional arrangements	Politics and decision-making	Conflicts and tensions — Limiting factors	Conflicts and tensions — Principal interest to benefit
Regulative planning	Local authorities	Technical-political	Strength of local market	Local land and property owners
Trend planning	Local authorities	Non-strategic gate-keeping	Retaining any control of market	Incoming developers
Popular planning	Community organisations	Imperfect pluralist	Community control of resources	Local lower-income groups
Leverage planning	Quasi-governmental agency	Corporatist	Potential for market revival	Incoming developers
Public investment planning	Quasi-governmental agency	Administrative corporatist	Long-term resource commitment	Local lower-income groups
Private management planning	Private trust	Paternalistic	Inability to move beyond tokenism	None

Source: Brindley et al. (1996: 159).

boundaries and scope for autonomy, difference and experimentation. Two significant consequences stand out. First, local planning authorities were increasingly encouraged to work across boundaries and tiers to create functional planning areas and multi-agency partnerships. The implications of different styles and cultures working alongside and with each other not only raise questions about how this is managed but also have implications for any institutionalist understanding. Second, Labour's growth agenda, which emerged from the *Sustainable Communities* plan (DTLR, 2002b), forced a step-change in partnership arrangements between authorities either imposing a new vehicle (e.g. an Urban Development Corporation) or 'encouraging' authorities to create their own, bespoke arrangements (see Chapter 5). This punctuated evolution did not arise through a process where discourses evolve until they no longer adequately explain or frame a situation, as is hinted at by New Institutionalism. Such shocks within planning arrangements were partly in response to the claim of Lowndes that change is the exception rather than the rule within local governance.

The existence, juxtaposition and fusion of planning cultures within such step-changes provide a further layer of complexity within New Institutionalist understandings of planning in the Labour era. Another dimension is that planning cultures and management cultures co-exist within organisations, as Schmidt (2008) highlights, though these are not necessarily aligned. What I would argue is that, given the particular circumstances of planning, both historically and professionally, the role of culture or styles has heightened significance with any institutional understanding of change during the Labour era. This significance rests on the exaggerated importance of professional discretion within planning, to which I turn below. However, Labour introduced its own cultures of planning through visions or imaginary places and processes, particularly around the notions of urban renaissance (Chapter 4) and spatial planning (Chapter 5). The urban renaissance agenda contained many positive images of largely middle-class living around what has been termed a 'cappuccino culture' (see Lees, 2003). Spatial planning, on the other hand, was portrayed as a conflict-free, coordinated and growth-friendly process for planning (Allmendinger and Haughton, 2009b). Both sought to impose themselves upon and jostle for attention with more local and longstanding cultures, images or styles with mixed success.

Discretion

New Institutionalism focuses attention upon the role of agents and their day-to-day practices of bargaining and negotiation within networks of interests. The significance of actors has long been understood within the policy implementation literature (e.g. Lipsky, 1980; Barrett and Fudge, 1981), and the daily, communicative practices of planners have become a major focus of research (e.g. Forester, 1989; Fischer and Forester, 1993; Healey, 1996). As planning cultures need to be understood within an institutionalist framework, so discretion in general and the changing nature of discretion under Labour also needs to be highlighted, particularly because discretion in planning is under-researched and – while this is acknowledged – little understood (though see Booth 1996, 2003, 2007). The UK system combines two approaches to government that make discretion an important element in planning practice. On the one hand there is the 'policy interpretive model', which posits a general guidance and oversight role in Westminster and allows local planning authorities to interpret this in the light of local circumstances and 'material considerations'. This gives local planning authorities considerable discretion and flexibility to determine both what is material and to what extent it is material in any particular case. Combined with a general reluctance of the courts to get involved following the Wednesbury decision in 1948 (*Associated Provincial Picture Houses Ltd* v. *Wednesbury Corporation* [1948] 1 KB 223), the upshot has been that local planning authorities have considerable discretion in the area of planning (Booth, 2007).

On the other hand, as Adler and Asquith (1993) point out, discretion in the UK has developed alongside powerful professions that lay claim to esoteric knowledge and control. This has carved out an influential role that encompasses both political *and* professional discretion (Booth, 1996). The former involves acting as an adviser or advocate to the authority in planning matters, while the latter is based upon judgement and values drawn from being members of a professional body. There are overlaps between the two, though it is the latter that is largely overlooked and has greater significance for institutionalist analysis.

During the Labour era, both political and professional discretion underwent changes as the government sought to circumscribe autonomy and enforce a more centrally directed policy agenda. Two specific initiatives stand out. First, the re-emphasis upon the plan-led approach through Section 38(6) of the 2004 Planning and Compulsory Purchase Act sought to reinforce the presumption originally introduced in 1991. Both provisions aimed to ensure that decisions were taken in accordance with the plan unless material considerations indicated otherwise. While there is disagreement over the difference in language between the two sections, the intentions are broadly the same. This raises the question of why the new wording was introduced. One possibility was to re-emphasise the continuum between planning policy and decision-taking under the new system of development planning introduced at the same time. As part of the new system the government also sought to fold in development control decisions by linking the two – a shift towards a more prescriptive approach reminiscent of zoning systems (Allmendinger, 2006).

Significantly, Labour began to undermine the plan-led approach particularly towards the end of the era when, following consideration of the issue in the second Barker report (Barker, 2006a), a draft PPS4 stated that, outside of town centres, planning applications for economic development-related activities should be considered 'favourably unless there is good reason to believe that the social, economic and/or environmental costs of development are likely to outweigh the benefits' (DCLG, 2009d: paragraph EC12.3). This 'top-down' mechanism to limit discretion was reinforced through the expansion in central government guidance. This expansion did not originate under Labour, though it was certainly accelerated. However, while some of the growth in national policy and guidance can be attributed to the expansion in issues and objectives falling under planning, there is little doubt that government also sought to increase central prescription in an attempt to improve the efficiency and effectiveness of planning to wider public policy objectives. Other central attempts to limit discretion and the autonomy of local areas were the shift towards regional housing targets, the introduction of binding inspector's reports on Local Development Frameworks, and the wholesale removal of the consideration of major infrastructure projects to the Infrastructure Planning Commission.

The second main attempt to limit the scope of discretion was through the introduction of performance indicators and targets linked to incentives in the form of the Planning Delivery Grant (see Chapter 6). Combined with the reinforcement of the plan-led approach, which aimed to ensure that decisions were taken in accordance with the plan and, ultimately, government policy, the introduction of performance targets and incentives sought to guarantee that such decisions were taken quickly and efficiently.

Labour's attempts at centralisation and the diminution of discretion were undermined by a range of factors. The re-emphasis upon the plan-led approach has been weakened by the lack of progress on the new system of development planning. Ironically, this has strengthened the significance of central policy but allowed local authorities greater scope to determine what is material within broad national-level regimes and out-of-date or premature local frameworks. The reasons for the lack of progress with the new frameworks were complex, though one significant factor was the shift of planning professionals to the growing private sector and to development control, which was being prioritised within authorities because of the impact of the Planning Delivery Grant. Another reason for the lack of progress concerned professional discretion itself: the absence of a local development framework maximised professional discretion (see Chapter 5).

Performance indicators and incentives led to widespread strategic behaviour and perverse outcomes, as local authorities refused applications rather than negotiated them (Allmendinger, 2009). Professional discretion was employed to determine the extent to which a proposal was acceptable in terms of policy objectives and impacts and whether and to what extent negotiations should be undertaken. Rather than limiting discretion, the slow progress of Local Development Frameworks combined with the growth in national policy that sought to achieve multiple, often ill-defined objectives actually increased local political and professional discretion. A further outcome was that, rather than strengthening the link between development plans and development control, the increase in local discretion focused greater attention on development control, with the result that it became increasingly politicised.

As opposed the deregulatory instincts of the New Right, the New Left sought to limit discretion by re-regulating the links between policy and decisions operating at different scales, centralising the policy framework and measuring and incentivising decisions. The outcomes demonstrate, among other factors, the continued importance of discretion within planning.

Conclusions

What this chapter has attempted to do is provide a meso-level framework to understand change in planning by emphasising the concerns of New Institutionalism with the political economy of neoliberalism. In doing so I am attempting to focus attention on the political economy of governance in order to re-emphasise the distributional and highly political dimensions of planning (Geddes, 2006: 78). Nevertheless, there is no simple way of understanding a complex, evolving and highly contingent state activity and professional sphere such as planning. The significance of the New Institutional understanding of change in planning becomes clearer from the experiences of the Labour era. However, such an understanding needs to emphasise aspects that are of particular pertinence to planning. The latter have been identified as the importance of planning cultures or doctrines and the continued influence of discretion. However, the chapters that follow will provide evidence to develop further any New Institutional understanding by pointing out particular elements of practice and change and addressing questions around the significance of agencies and structures in planning and how the multiple discourses under Labour helped shape knowledge and action within more local institutional rule sets. As such, this chapter is a starting point that focuses attention on the chapters that follow.

4

PLANNING AND URBAN POLICY

Introduction

Urban policy under New Labour was a diffuse, complex and evolving affair. Any assessment will depend largely on what is taken to be the scope of the field. 'Wide' or permissive definitions rightly emphasise the diversity of urban policy and highlight the multiple initiatives and cross-cutting, issue-based and (sometimes) contradictory approaches (e.g. Johnstone and Whitehead, 2004). More discrete evaluations focus instead on particular dimensions of urban policy or neighbourhood renewal, such as Health Action Zones (e.g. Barnes *et al.*, 2005). Others are case studies exploring the 'joined-up' policy approach of New Labour (e.g. Raco and Henderson, 2009) or focus upon the new vehicles such as Urban Development Corporations (e.g. Imrie and Thomas, 1999) or new strategies such as the Northern Way or Sustainable Communities Plan (e.g. Gonzalez, 2006). Yet others explore specific schemes, such as the Housing Market Renewal Pathfinders (e.g. Bramley *et al.*, 2007), approaches to finance, including Business Improvement Districts (e.g. Peel *et al.*, 2009), Tax Increment Financing, or attempts to link create multi-scalar coordination through Local Area and Multi-Area Agreements (e.g. Morphet, 2008). And then there are those who focus upon outcomes for specific groups such as the young (e.g. Brown and Lees, 2009) or the homeless. The scope of what falls within what we could broadly term urban policy is considerable, especially as the boundary between 'rural' and 'urban' issues has become increasingly blurred as attention has been drawn to common themes such as accessibility and climate change.

The wide scope of urban policy under Labour is matched by its complexity. Despite the claims of Hall (2003) that New Labour's approach to regeneration and neighbourhood renewal is coherent and structured, the majority view highlights tensions and contradictions. Colomb (2007) argues that one way to make sense of Labour's urban policy is to distinguish between, on the one hand, a neighbourhood renewal stream, which focused upon tackling social exclusion at the local level, and, on the other, a redevelopment stream, relabelled an 'urban renaissance', which concentrated upon the economic, physical and cultural/social regeneration of towns and cities. Others analyse Labour's approach in different ways, emphasising and unpicking the role of community and inclusion (e.g. Imrie and Raco, 2003), the suitability of current governance arrangements (e.g. Brownill and Carpenter, 2007; Allmendinger and Haughton, 2009a, 2007), the creation of new narratives around the notion of 'urban' (e.g. Lees, 2003; Hoskins and Tallon, 2004), the role of community involvement (e.g. Imrie, 2009) and the privatisation of public space (Minton, 2009).

This chapter focuses primarily upon the role of planning in urban policy which, although central, has been largely overlooked. According to Healey (2004b), the approach from 1997 was an attempt to link urban policy to land-use planning and development in an effort to move away from a focus upon 'problem areas' to the qualities of cities and places. Planning was a key element in the New Labour model, which depended, in large part, upon a private-sector, development-led approach to its aim of an 'urban renaissance'. As opposed to the deregulated and *de minimis* role for planning in the early New Right's property-led redevelopment approach, typified by the experiences of London Docklands, New Labour's vision was one of active public–private partnerships around a broad policy framework of 'urban renaissance' and 'sustainable development'. There was a widespread assumption that planning was a relatively uncontentious component of urban policy: Labour assumed that its urban renaissance agenda and the notion of sustainable communities fitted within the aspirations of local communities and the planners and others charged with delivery.

It is difficult to disagree at a broad level about the thrust of Lord Rogers's desire (DETR, 1999), echoed in the Urban White Paper (DETR, 2000b), for an 'urban renaissance' involving liveable, inclusive, high-quality environments. In many cases, however, the physical nature of urban policy interfaces with broader sceptical attitudes towards development within communities, as well as the need to make sense of rather abstract themes and concepts such as 'sustainable communities' in relation to actual and locally anchored development schemes. In other words, any implementation of urban policy demonstrated a naivety about the role of planning and the nature of change, since its elements could not be 'read off' from multiple and (necessarily) vague aspirations. This issue was one that plagued other areas of planning (see Chapter 5 on spatial planning, for example). A second and related theme in this chapter concerns the competing objectives of policy and how their resolution is left largely to localities in general and the planning arena in particular. In some ways there is nothing new in the degree of reconciliation of such tensions. However, under New Labour the miasmic and ambiguous national policy framework had significant implications when combined with the desire to deliver a step-change in urban development generally and housing development in particular. Planning at the site level, where urban policy is 'made real', became a battleground for competing views around the direction of urban policy (see, for example, Brownill, 2007). This need to reconcile disparate aims was evident at a local, 'everyday' level of planning as well as in flagship initiatives such as the Thames Gateway and the Olympics.

Labour and urban policy

One way to conceive of urban policy during the first decade of the twenty-first century is to identify four broad eras, each of which was characterised by different emphases upon renewal, regeneration and redevelopment. As with planning generally, the early, post-1997 period was epitomised by a continuation of the policies of the Major administrations. In opposition the Labour Party had committed itself to the spending plans of the Conservatives, and it continued with the renewal-dominated regime of initiatives such as the Single Regeneration Budget. Nevertheless, this early phase was short lived, and John Prescott, who had responsibility for the key elements of urban policy in the Department of the Environment, Transport and Regions and then the Office of the Deputy Prime Minister, complemented the renewal emphasis of government policy with a more holistic regeneration stream through the establishment of the Urban Task Force (DETR, 1999a) and the Commission for Architecture and the Built Environment, while also overseeing the Urban White Paper (DETR, 2000b). Key to this second period were the notions of design and sustainability, or creating places where people wanted to live (again) – or, as Hoskins and Tallon (2004) put it, creating an image of the 'urban idyll' to attract the middle classes back to the cities. This 'urban renaissance' was broadly welcomed within the profession and beyond, reinvigorating the role of planning and design, linking previously disparate elements and sectors into integrated 'place-making' and highlighting the role of social inclusion (Healey, 2004b). However, in parallel to the Prescott and ODPM emphasis upon creating high-quality urban environments was the Treasury-driven concern with economic competitiveness, including the role of housing affordability (Allmendinger and Tewdwr-Jones, 2009). The two strands of policy existed uneasily in places, particularly around the notion of significant housing growth in new settlements and the implications for the green belt (CPRE, 2006).

The publication of the government's *Sustainable Communities* programme in 2003 heralded the initiation of a third phase, which sought to amalgamate and reconcile these themes. The programme announced a step-change in housing provision concentrating on four 'growth areas' (ODPM, 2003a). The locations were in the high-demand areas of the south of England, and the emphasis would be upon creating 'sustainable communities'. Among further changes was the large-scale demolition and 'market renewal' of housing in nine areas of the Midlands and the north of England announced in 2002. Such concerns over housing delivery were highlighted in the first Barker report (Barker, 2004), which helped complement the regeneration theme of urban policy with a reminder that both development *and* redevelopment would be necessary parts of housing improvement. While the latter could incorporate some of the principles of sustainable communities, however vague the notion, the emphasis, from the Treasury, was clearly upon housing numbers and supply. The role of bodies such as Regional Development Agencies and English Partnerships became more explicitly concerned with delivering development, regardless of whether such development amounted to regeneration.

With the onset of the credit crunch in 2007 and the subsequent recession, regeneration took a much clearer economic emphasis in the fourth phase of urban policy (DCLG, 2009a). The introduction of Local Economic Assessments and the merging of regional economic and spatial strategies into a single integrated strategy prepared by the Regional Development Agency were heralded by the Treasury's *Review of Sub-National Economic Development and Regeneration* (HM Treasury, 2007a). Further shifts toward an emphasis upon development and delivery came with the decision to move ahead with the Infrastructure Planning Commission, the review of Regional Spatial Strategy housing targets announced in the Housing Green Paper (DCLG, 2007b), and the eco-town initiative, which sought to bypass development plans and kick-start housing development, albeit based around sustainable development criteria.

Clearly, then, there were evolving themes in New Labour's urban policy. It did provide a distinct break with the previous regimes in its understanding of the issues and role of cities as more than reflections of economic activity, in its aims for cities (particularly the desire to move towards creating 'cities for people') and in the mechanisms through which urban policy would be delivered. However, such intentions were not matched by a coherent approach or consensus within government. The shift was probably best illustrated in the draft Planning Policy Statement 4, published in late 2007. While the 2004 Planning and Compulsory Purchase Act had reinforced the 'presumption in favour of the development plan', draft PPS4 muddied the waters when it came to proposals that involved economic development. When determining applications, local authorities should 'consider proposals favourably unless there is good reason to believe that the economic, social and/or environmental costs of development are likely to outweigh the benefits' (DCLG, 2007h: paragraph 29). The changing emphasis of urban policy under Labour highlighted the competing objectives and perceptions of roles for planning within government which amounted to five distinct though simultaneous themes.

Economic competitiveness and neoliberalism

There is a strong body of opinion that current urban policy is a form of 'rolled-out' neoliberalism (Peck and Tickell, 2002) and that planning policy and the role of planning in urban policy is to serve the interests of capital and further facilitate the success of financial services in London (Massey, 2007; see also Jones and Ward, 2004; Raco, 2005). Advocates of this position assert that, in contrast to the Thatcherite 'rolled-up' neoliberalism, which saw a minimalist role for the state, urban policy under Labour was about providing the infrastructure and social capital to support to the market. Such views are simplistic and rely, to an extent, upon caricatures and generalisations. While initiatives such as Urban Development Corporations in the 1980s were actually far more interventionist than is widely acknowledged (Allmendinger and Thomas, 1998), Raco (2005) rightly argues that the notion of 'rolled-out' neoliberalism is a rather crude characterisation of urban policy under New Labour that also includes other elements and rationalities. Nevertheless, there is little doubt that the focus of regeneration, driven largely by the credit crunch, became unashamedly economic: 'successful regeneration strengthens communities by creating new economic opportunities' (DCLG, 2009a: paragraph 4). The government proposed greater targeting of scarce resources to areas 'where there are opportunities for transforming the economic prospects of areas with lower economic performance' (ibid.: paragraph 17). The credit crunch and the recession that emerged in 2008 clearly focused the minds of government ministers on priorities, though, 'despite the rhetorical promotion of the concept of "sustainability", the belief that commercial, market-led success represents the basis for new urban spaces still dominates agendas' (Raco, 2005: 56).

The significance of economic development and competitiveness under New Labour was witnessed by the well-funded and explicit role of Regional Development Agencies as proactive and interventionist. As Colomb puts it 'It [Labour's urban renaissance agenda] can be interpreted both as an attempt to deal with, and address, the adverse impacts of neoliberal political and economic restructuring on the inner city, whilst being part and parcel of the neoliberal urban project' (2007: 17).

Design and culture

A significant theme of urban policy under Labour was the reimaging of the city and the creation of attractive, safe and stimulating places (Kearns, 2003). This theme had a number of strands. At one level, the Rogers report and subsequent Urban White Paper sought a semantic reimaging of the city to create what Hoskins and Tallon term the new 'urban idyll' or order, to 'promote a vision of town and city-centre living as a desirable alternative to suburban and rural life' (2004: 25). This was attempted through discourses around urban life as the source of economic and social well-being, cultural mixture and citizenship, including comparisons with successful urban living in places such as Barcelona. The key to the urban renaissance was therefore a design- and culture-led strategy which aimed to attract young, urban professionals with large disposable incomes and high social, educational and cultural capital back to urban living (Colomb, 2007: 8). A second strand to this theme was physical and involved the promotion of 'good design' to encourage 'civilised' and 'civilising' behaviour and foster a sense of community. Related to the social inclusion and community theme below, the creation of new spaces had a direct impact upon behaviour through reducing motivation towards uncivilised behaviour (e.g. mixed-use developments, CCTV) and through inclusion in the process of place creation, which sought to foster citizenship, identity and pride (Imrie *et al.*, 2009). Third, local services formed a critical component in the strategy of place-making, with Labour focusing particular attention upon day-to-day service delivery and the process and governance of localities (Morphet, 2008).

According to Raco (2005), place-making and distinctiveness in regeneration have, in reality, focused more upon consumerism. Mixed-use redevelopments are typically retail-led and anchor major brands and retailers into schemes, creating, at the very least, a tension with the notions of distinctiveness (2005: 55–6).

Joined-up governance

The call for more integration in urban policy and the need to overcome traditional, professional and administrative boundaries was not new to Labour (see, for example, Robson *et al.*, 1994). What marked Labour out was the determination to pursue coordination and integration in urban policy (Thornley and West, 2004; SEU, 1998). A range of initiatives and reports were produced to ensure that the wide variety of schemes that came under the umbrella of urban policy were co-ordinated (Tiesdell and Allmendinger, 2001; Lawless, 2004). The ubiquitous notion of partnership was also used as a tool through which such coordination on the ground could be achieved (Davies, 2001). Local Area and Multi-Area Agreements provided a key mechanism for coordinating urban policy, with the emerging regional strategy providing a further sub-regional vehicle (DCLG, 2009a: 19). These built upon the perceived success of Urban Regeneration Companies, established through the partnership of local authorities, Regional Development Agencies and other business and community stakeholders while also seeking to reinvigorate the less than successful Local Strategic Partnerships (LSPs). The LSPs, in particular, suffered from being public-sector dominated and struggled to reconcile the tension between coordination, with the necessary lengthy timescales involved, and increasingly target-led approaches, which inevitably favour shorter-term timescales.

There was a gradual realisation that joined-up governance in general and partnership-based approaches in particular were not achieving their aims, were becoming a habitual and automatic response to funders' requirements rather than a better means of delivering urban policy ends (Parkinson *et al.*, 2005) and were leading to 'partnership fatigue' among participants (Hastings, 2003). As Cochrane (2007) notes, it is easy to forget the difficulties in achieving a coordinated approach, and Simmons (2009) points to how, even when there is a commitment, outcomes are often less than coordinated for what might be basic disagreements. In their study of the role of the London mayor in providing coordination and integration in the urban and spatial policy across the capital, Thornley and West (2004) found that political priorities remained a significant barrier.

Social inclusion and community

A key aspiration of urban policy and the urban renaissance was the need to tackle social, political and economic exclusion (SEU, 2001). While there were a range of initiatives aimed at this, mainly through the Neighbourhood Renewal Unit, the role of planning in tackling exclusion has been more ambiguous. Lees *et al.* (2008) argue that a key component of urban policy was the need to gentrify urban areas, allowing the middle classes to mobilise resources and benefit from rises in property values. Such gentrification also required the 'sanitisation' of urban areas through the removal of antisocial behaviour (both disruptive and criminal) and the management of urban space (see MacLeod, 2002; Atkinson, 2004; Colomb, 2007). This amounted to what Whitehead (2004) characterises as new spaces of morality which reflect the themes within Third Way thinking around civic responsibility.

As discussed above, the aim of creating mixed communities and tackling social exclusion came up against the perceived need to generate attractive places for the middle classes that are safe and homogeneous:

> The repackaging of city centre neighbourhoods as chic lifestyle choices for young, middle-class professionals has done little to advance social inclusion. Instead, those who do not fit the image of the urban idyll, who seem 'out of place' due to their behaviour, apparent status, dress or age, are being gradually purged from the streets of these solipsistic enclaves.
>
> (Johnstone, 2004: 87)

Whether the lack of tolerance to begging, rough sleeping and public drinking in city centres amounted to social exclusion, as Johnstone claims, is, however, a moot point. Nevertheless, there was undoubtedly a tension between initiatives such as the 'respect' agenda, aimed largely at the young (notably the focus on antisocial behaviour), on the one hand, and the wider rhetoric of social mix and inclusion, on the other. The result was a displacement of certain social groups, including the homeless, the young and beggars, from 'regenerated' places (MacLeod, 2002).

Climate change, housing and the environment

What is often overlooked in analyses of urban policy is the significance of the rural and the wider role of planning in creating the policy context for regeneration through affecting land and property markets. A related theme concerned the pressures to deliver increased housing supply while doing so in a way that helps mitigate the impacts of climate change. Such challenges fall squarely within the planning field.

The need to increase housing supply in order to affect and improve housing affordability was widely accepted within government (Barker, 2004, 2006b; DCLG, 2007a), though less so within the communities where such housing is supposed to be located (NHPAU, 2009). The aspiration of delivering 2 million new homes by 2016 and 3 million by 2020 (DCLG, 2007b) led to successive changes in the policy and advice from government offices to local authorities preparing strategies, which had to be balanced with the avalanche of advice on other issues and objectives. One inevitable outcome was a form of 'analysis paralysis', as local planning authorities and others sought to balance competing and sometimes irreconcilable demands. However, the notion that Regional Spatial Strategies would provide apolitical, aspatial housing targets that would 'flow' to those charged with preparing Local Development Frameworks proved naive – a point partially recognised in the proposed changes to create single integrated strategies and the proposal by the Conservative Party to abolish the regional tier of planning entirely (HM Government, 2010). Regional Assemblies were partly composed of councillors who represented the very places that would be tasked with finding sites. Despite government office insistence and the necessary evidence base, some Regional Assemblies did not accept officer advice on growth in general or housing numbers in particular. As Zetter puts it:

> In the regrettable absence of a national spatial plan and given the lack of an intermediate level of government to arbitrate between national and local interests, 'not in my back yard' tendencies are encouraged because the larger picture is missing.
>
> *(Zetter, 2009: 262)*

The pressure to deliver more housing and tackle worsening affordability coincided with other policy objectives around the environment and climate change. On top of the longstanding commitments to protect the countryside from development for its own sake and to limit urban sprawl, two further policy objectives emerged. The first was a wider environmental concern linked to biodiversity and natural assets, such as air and water quality, that emphasised the need for new development to live within 'environmental limits' (ODPM, 2005a). What this meant in practice was unclear, particularly as vague terms such as 'unacceptable impacts' provided considerable scope for interpretation. Planning and local authorities had limited experience in using such approaches, and a lack of data provided another hurdle (Smith and Pearson, 2008). Further, there was a clear though unacknowledged limit to the extent to which planning could address environmental concerns, given the lack of tools to influence implementation and management of development. The second policy objective concerned the role of planning in mitigation and adaptation to climate change. Labour promised to put the environment at the 'heart of policy-making', though it struggled to translate global concerns and responsibilities into local action and outcomes, particularly as local framing of environmental issues allowed different interpretations to co-exist (Wilson, 2009). The justified hyperbole around the threat of climate change (see, for example, DCLG, 2010e) was not matched by advice and policy, creating a hiatus that was filled by a variety of interpretations. More significantly, the context of climate change helped shift the onus of proof through the requirements of supporting information to applicants and developers to demonstrate the sustainability of their schemes.

The combination of an underlying resistance to greenfield development and a powerful though vague new policy stream around the environment and climate change further prioritised the need to focus upon previously developed land and increased development densities (DETR, 2000a; DCLG, 2006j). Combined with the growing importance of development-related delivery of affordable housing and the lack of clarity around climate change, sustainability and the environment, urban policy began to become overloaded with a spectrum of aspirations.

The critical and largely overlooked dimension to urban policy under New Labour was the role in planning in reconciling such disparate themes. While it is almost a truism to say that the themes of urban policy under New Labour were, at best, in tension and, at worst, in contradiction, as Cochrane (2007) rightly points out, this is nothing new for urban policy in the UK. Brownill (2007: 264) argues that these tensions are held together and reconciled at a national level through discourses such as 'urban renaissance' and 'sustainable communities', leaving the local and other tiers to reconcile and interpret them. This echoed the New Localism of Labour and the view that government should provide communities with the means to develop local solutions (SEU, 2001: 2).

As a consequence, the role of planning in urban policy had two main dimensions. There was a positive dimension that allocated planning a role in both the process (e.g. coordination and integration) and the outcome (e.g. sustainable communities or urban renaissance). However, there was also an equally important negative or control dimension, which required a more traditional control of development – for example, to limit urban sprawl, increase land prices by restricting supply and control out-of-town retail developments. 'Place-making' in its widest sense involved the need to inhibit as well as facilitate. This dual role, including the more creative and positive function for planning in urban policy and the necessary control dimension, was initially overlooked. It is the latter in particular on which competing discourses of and problems with urban policy were focused.

Urban policy in practice

In order to highlight and explore these points, I now turn to two contrasting examples of planning and urban policy during the Labour years. They are not untypical and represent the nature and difficulty of delivering urban policy and the pressure on local planning authorities and communities to resolve competing interpretations of sustainable development, urban regeneration and the role of planning. The first is the regeneration of the area to the north of King's Cross station in London, while the second is the redevelopment of a large area of previously used land in Radstock, a small market town in Somerset. The scale and cost of the schemes are poles apart, though the issues around the role of planning in delivering urban policy and regeneration are strikingly similar.

King's Cross

Works on the redevelopment of the area to the north of King's Cross station, a 67 acre site with a planned 1,900 homes and 8 million square feet of mixed uses, began in 2007. The developer, Argent, estimates the scheme will not be complete until 2020. Views on when the process began differ, though a reasonable starting point would be the adoption in 2000 of the Camden Unitary Development Plan, which identified the King's Cross Opportunity Area for a mixed-use development of employment and housing, including affordable housing. The desire to redevelop the area goes back much further, however (Fainstein, 1994). It is difficult to disagree with Argent's view, endorsed by both the London Plan and the planning inspector for the 'Triangle Site' appeal (see below), that:

> The King's Cross scheme is probably the most important regeneration project currently in progress in Central London. It is a scheme of immense potential benefit in environmental, social and economic terms. It has emerged against a background of chronic dereliction and urban decay. For more than 30 years efforts have been made to stimulate and secure the urban renaissance of this part of the metropolis.
> *(Planning Inspectorate Case APP/X5210/A/07/2051898: paragraph 6.1)*

The development is of national significance given its proximity to both King's Cross and St Pancras railway stations, the latter being the Channel Tunnel rail link terminus. The scheme amounts to 'the biggest inner city redevelopment in Europe' (Holgersen and Haarstad, 2009: 348).

The development area falls within the London Borough of Camden (LBC) (64.5 acres) and the London Borough of Islington (2.5 acres). Despite the large area of land that was previously railway related, there is an existing community of residents and local businesses (Map 4.1).

In line with national planning guidance on the need for 'front-loading' public consultation and the view of both planning authorities involved of the need for a partnership-led approach, extensive consultation was undertaken prior to the submission of planning applications. Between July 2001 and June 2003, representatives from over 150 community groups and 4,000 individuals were involved. A bespoke group, the King's Cross Development Forum, was established by LBC as an umbrella organisation for 160 community groups to provide a focus for discussions over the redevelopment. Further consultation, including 30,000 mailings to local people and businesses, consultation workshops and public exhibitions, was undertaken by LBC once the planning applications had been submitted. This process was repeated when the proposals were amended following the response from the first round of consultation. Argent has won numerous national awards for consultation on King's Cross, and the overall approach has been characterised as 'positive' (Imrie, 2009: 106). Planning permission for the redevelopment was granted by LBC in December 2006. Table 4.1 sets out the key stages in the redevelopment scheme.

Notwithstanding the above, to say that the scheme has been controversial would be an understatement. A small group of local residents led a determined campaign to thwart the scheme: 'We all want regeneration but this huge scheme with its bleak office blocks is not what the area needs. It's designed to make money

72 Planning and urban policy

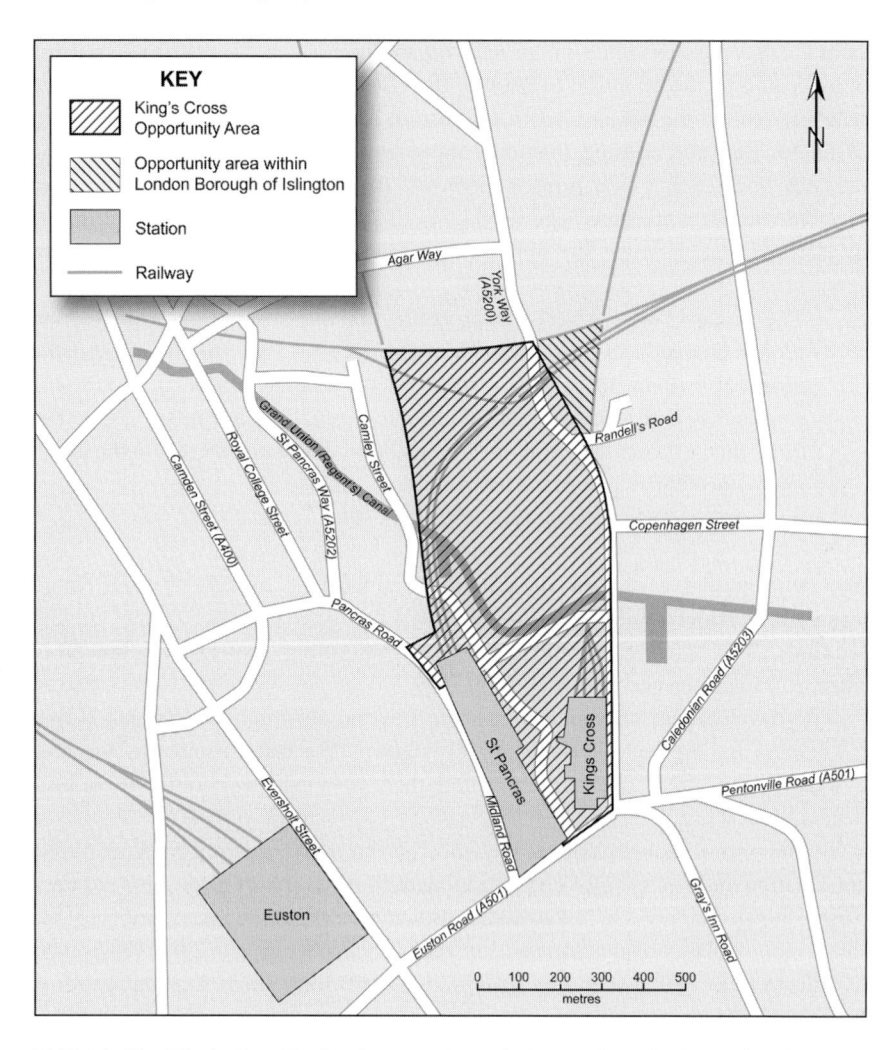

MAP 4.1 The King's Cross Redevelopment Area: the 'Triangle Site' is located in the top right-hand corner and hatched as 'Opportunity Area within the London Borough of Islington'.

not to help the local community. We want the council to think again' (King's Cross Railway Lands Group, 2006). This is despite the redevelopment of the area complying with national, regional and local planning frameworks and the evolution of the design and layout being undertaken in close collaboration with the two London boroughs. There was also considerable local support from groups and individuals. Despite the overwhelmingly supportive policy context, a critical element in any redevelopment scheme is the need to secure planning permission.

Two incidences stand out in the struggle to secure planning permission. The first concerns the 'Triangle Site', or the 2.5 acre element of the scheme that falls

TABLE 4.1 Selected key events in King's Cross redevelopment

Date	Event
2000	Argent appointed.
March 2000	Camden Unitary Development Plan (UDP) adopted, identifying the area for redevelopment and setting out objectives.
July 2001	Argent publishes *Principles for a Human City* consultative regeneration framework document, which stated that the objective for King's Cross is to devise and then deliver over the next fifteen or so years an exciting and successful mixed-use development – one that will shape a dense, vibrant and distinctive quarter, bring local benefits and make a lasting contribution to London.
Oct 2001	Camden publishes consultation *King's Cross: Towards an Integrated City*, including key objectives, and endorses Argent's *Principles for a Human City* as a positive and enlightening statement of urban policy for the twenty-first century.
Dec 2001	Camden publishes a deposit draft of a revision to Chapter 13 of the UDP, which designates the site an 'Opportunity Area' for regeneration.
Dec 2001	Argent publishes the consultation document *Parameters for Regeneration*, identifying eighteen parameters for regeneration, including taking forward Camden's vision.
June 2002	Draft London Plan published identifying King's Cross as an 'Opportunity Area' and recognising the importance of development for the capital.
June 2002	Camden publishes *King's Cross – Camden's Vision*, which refers to Argent's framework as containing good ideas on how to make King's Cross central and surrounding areas a better place in which to live, work and travel. It welcomed the commitment to providing good-quality design, the importance given to the street environment, and the recognition of the unique qualities and sense of place created by the historic buildings.
July 2002	Inspector endorses Camden UDP 'Chapter 13' approach to King's Cross redevelopment.
Sept 2002	Argent publishes consultation *A Framework for Regeneration*, which built on the ideas and information developed in the preceding consultation documents. The document described an emerging framework of new public routes and places, presenting a range of development ideas for each part of the proposed framework for comment.
May 2003	Camden UDP 'Chapter 13' adopted.
June 2003	Argent publishes *Framework Findings*, which includes an analysis of responses to the consultation documents and process. *Framework Findings* was prepared in collaboration with consultation specialists FLUID, who helped to shape and manage the consultation process and analyse the results.
Dec 2003	Joint Camden and Islington Planning and Development Brief for King's Cross published. Brief sets out councils' objective of wishing to 'see major development and regeneration started, and completed, as soon as possible, to overcome the problems and uncertainties that have blighted this site in the recent past.'

Continued overleaf

TABLE 4.1 *Continued*

Date	Event
Feb 2004	London Plan adopted. Endorses Opportunity Area policy and highlights the need for high-density business development and housing.
May 2004	Three planning applications and eight heritage applications submitted.
May 2004	Camden publishes revised deposit draft of UDP.
July 2005	Draft Sub-regional Planning Framework published. Endorses 'Opportunity Area' approach.
Sept 2005	Three revised planning applications submitted following responses from English Heritage, the GLA and other interested parties.
Mar 2006	Camden resolves to grant planning permission.
April 2006	Islington resolves to grant planning permission.
Nov 2006	Camden signs Section 106 Agreement and issues consents.
Dec 2006	King's Cross Railway Lands Group applies for a judicial review.
May 2007	Judicial review hearing. Dismissed.
June 2007	Enabling works begin on site.
July 2007	Islington refuses planning permission for 'Triangle Site'.
July 2008	Appeal against refusal of planning permission upheld by secretary of state.

within the London Borough of Islington (see Map 4.1). Despite previously resolving to grant permission for the scheme, which conformed to the development plan, and against officer and legal advice, the authority refused permission in July 2007. Elected members determining the application were lobbied by a small group of community activists who were opposed to the whole scheme and saw the opportunity of halting this part of the development as a chance to achieve their aim (Evans, 2008: 94). The reason for refusal hung on the extent of affordable housing provision. Across the scheme as a whole, the development would lead to 1,946 dwellings, of which 834 would be affordable (44 per cent provision). Islington members' main objection was that, in the area of the scheme that fell within their borough, the proportion would be 34 per cent and therefore insufficient. This amounted to a U-turn on the part of Islington, which had previously accepted that the site be treated as a whole, particularly as the area that came under its control amounted to only 2.5 of the 67 acre total. This was the position it had taken in its negotiations and the joint development brief. As LBC put it in its evidence to the subsequent inquiry against refusal of planning permission:

> One of the core purposes of the development plan (and related Supplementary Planning Guidance, such as the Joint Brief) is to provide certainty for developers. The Appellants have been encouraged by both Councils to bring their proposals forward in good faith accordingly. In these circumstances it is surprising that Islington's evidence has focussed on figures for the

Triangle Site alone, has not referred to the different stance now being adopted, or provided any explanation for it. This is a complete departure from Islington's position during officer level discussions and from everything that had gone before.

(Planning Inspectorate decision letter for 'Triangle Site', case APP/X5210/A/07/2051898: paragraph 7.2)

The public inquiry explored this issue, with the inspector coming to the conclusion:

It seems to me that the developers have gone to considerable effort to discuss the appropriate housing mix with the Councils and the local communities. While significant differences remain with the latter, I consider that the developers have produced an exemplary range of provision that, especially with regard to the intermediate housing sector, fully reflects the principles advanced in PPS3 and its daughter document Delivering Affordable Housing.

(Ibid.: paragraph 12.59)

To nobody's surprise, but to the disappointment of a minority, the secretary of state upheld the appeal and granted permission. Ironically, because Argent had appealed against the refusal of its original application, Islington lost the changes it had negotiated, as the inspector approved the application as submitted. The outcome was less affordable housing than had been negotiated.

The second incidence was the judicial review brought by the King's Cross Railway Lands Group challenging LBC's decision to approve permission for the main site in February 2006. The basis of the action was that LBC had a duty to reconsider its decision to approve the application following the publication of revised government guidance on affordable housing in Planning Policy Statement 3 (published in November 2006). The judicial review was brought on the basis that, following the local government elections in May 2006 and a change in composition of the planning committee to that which had granted permission for the scheme in March 2006, officers incorrectly advised the new committee that they could not reconsider the provisional consent. Further, that the change in the definition of affordable housing in the new PPS3 was a material change in circumstances that should have led the committee to reconsider its previous decision. In his judgment, Mr Justice Sullivan described the officer's advice to the committee as 'impeccable' (*King's Cross Railway Lands Group* v. *London Borough of Camden*, EWHC 1515 [2007] (Admin): paragraph 76), and the application to quash Camden's decision was rejected.

The experiences of the King's Cross scheme highlight a range of issues around New Labour's approach to urban regeneration. First, the redevelopment of the area met the requirements of the vague discourses of urban renaissance, economic competitiveness, social inclusion, climate change and sustainability. Such notions were superficially attractive at a general level and helped create a consensus around the scheme at different spatial scales. The developers were able to demonstrate that their scheme met the necessary criteria through the deployment of numerous studies and other supporting information. However, opposition arose when the implications of what was meant by 'sustainable development' and 'urban renaissance' became apparent. The King's Cross Railway Lands Group had a very different notion of sustainable development based on low-cost housing, green spaces and community facilities. The commitment to consensus-, partnership-based regeneration in King's Cross was challenged by a commitment and strategy more akin to 'older style' confrontational politics (Holgerson and Haarstad, 2009). The implications were to challenge the idea that planning was a tool to manage growth rather than reconcile normative differences and act in the 'public interest'.

Second, the public inquiry and judicial review cost around £4 million and the development control (as opposed to planning) process took four years. One could, perhaps, understand better the determination of those who wished the scheme to fail had they not had adequate opportunity to make their views known or had the scheme been contrary to planning policy. Yet, clearly, the King's Cross scheme was progressed through the planning system. Section 38(6) of the 2004 Planning and Compulsory Purchase Act replaced Section 54 of the 1990 Town and Country Planning Act, though, in effect, both sought to achieve the same end of a 'plan-led approach'. In short, planning decisions should be made in accordance with the development plan unless there are good reasons to depart from it. The King's Cross scheme was developed in conjunction with emerging local, regional and national planning guidance. Against the overwhelming advantages of the scheme, the objectors did not want to see the area 'gentrified': 'If a colony of frogs in the Railway Lands gets displaced, the developers have a duty to find an alternative habitat for them, but there is nothing like that that applies to people' (King's Cross Railway Lands Group, in *Local Government Chronicle*, 15 February 2007: 20).

The Railway Group wanted a doubling of the 44 per cent affordable housing provision offered, which according to the developers would have made the scheme unviable and unlikely to proceed. As the inspector for the appeal against Islington's refusal of planning permission put it when granting permission for the triangle site: 'the Amenity Groups' concerns overlook the purposes of regeneration and its intended "ripple-out" effects to the wider area' (Planning Inspectorate Case APP/X5210/A/07/2051898: paragraph 12.27).

Some may argue that this drawn-out, costly and risky process is a reflection of the complexity of the scheme and its location. However, such experiences are also reflected in less high profile schemes.

Norton Radstock

Located 9 miles to the south of Bath, Radstock was the centre for coal mining in north Somerset from the late eighteenth century until the early 1970s. The Great Western Railway established a goods yard and station in the centre of the town in 1854, but the railway has remained unused since the early 1980s. The decline of coal mining echoed the economic decline in Radstock. In 1998, Single

Regeneration Budget funding was secured and the Norton Radstock Regeneration Company (NRR) was established by the local authority as a 'not for profit' company to take forward regeneration. NRR's directors were drawn from the local community and included some elected councillors from the local planning authority, Bath and North East Somerset (BaNES). A Community Planning Weekend held in 1998 identified the redevelopment of the disused railway land as a priority, and the land was purchased by NRR with help from the South West Regional Development Agency in 2001 (Map 4.2). The site is around 18 acres. During 2003, NRR undertook a master planning exercise to meet community aspirations. After competitive tendering, Bellway Homes was chosen as the partner to deliver the scheme and worked with NRR to develop a vision for the site. The original scheme comprised up to 210 residential units, up to 695 square metres of retail floorspace, up to 325 square metres of business floorspace, and the adaptation of some existing buildings on the site to provide 145 square metres for a heritage and community centre.

The site was allocated in the BaNES Local Plan for a comprehensive mixed-use

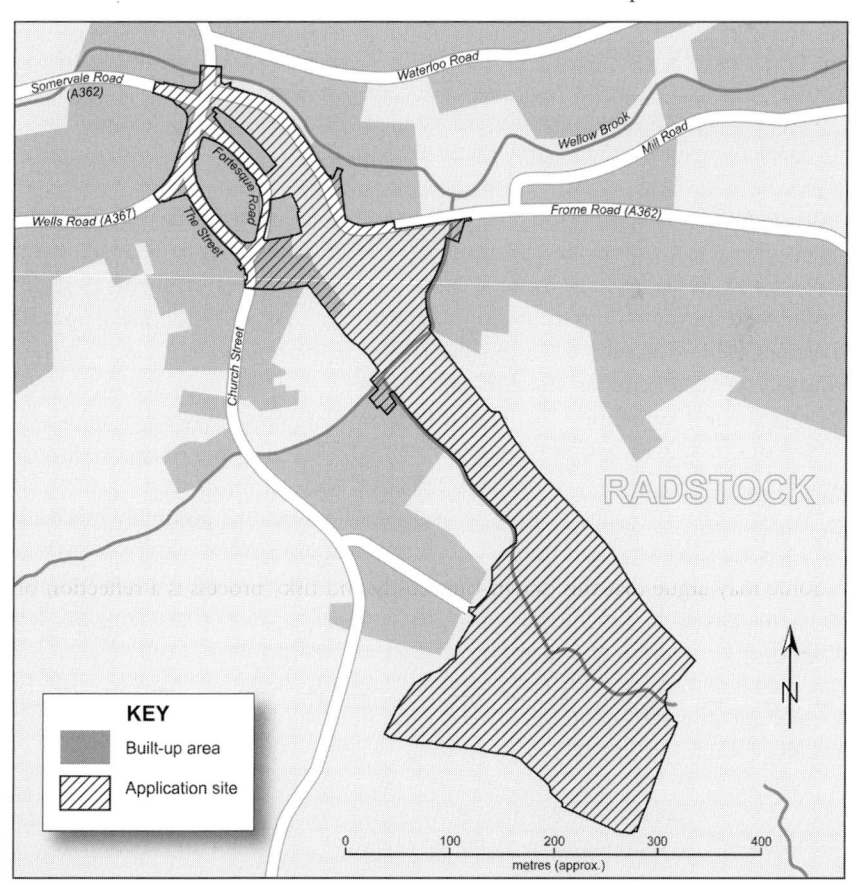

MAP 4.2 The Norton Radstock Regeneration Area.

scheme. The inspector's report for the local inquiry into the plan in 2004 recommended that any redevelopment of the site had to be approached in such a way as to ensure that its natural heritage was fully protected and that the housing element needed to be phased. The master plan for the site was completed in 2006 and public consultation was undertaken. In the local elections of May 2007 control of the council shifted from the Liberal Democrats to the Conservatives. The latter did not have a majority and so needed the support of independent councillors, two of whom had been elected for Radstock on the issue of stopping the development.

In March 2007 BaNES voted in principle to support the scheme, subject to a Section 106 Agreement, and following negotiations outline permission was issued in March 2008. The Section 106 Agreement covered the:

- preservation and management of railway features to be retained;
- implementation of an Ecological Mitigation Compensation and Management Plan;
- provision and management of public open space and play areas;
- implementation of the highway works;
- provision of a new town-centre car park;
- provision of a cycle route through the site;
- provision of affordable housing within the development;
- implementation of a local employment charter.

In 2008 the reserved matters application was submitted for the first phase of the development, which included 83 dwellings (of which 35 were affordable). As at King's Cross, those opposed to the scheme had lobbied elected members to revisit the approval of the outline application. The officer's report to the Development Control Committee, which was to consider the application on 8 July 2009, stated: 'Members are advised that this does not represent an opportunity for the public, or indeed the Committee, to revisit the principles of the development which were permitted as part of the Outline proposal' (BaNES Development Control Committee, Application Number 08/02332/RES: 1).

Also echoing the King's Cross case, objectors explored the possibility of a judicial review of the council's original decision. The prospect of a review was one factor which led the Development Control Committee to defer the proposal for three months while the council considered its legal position. Once approval was forthcoming, a judicial review was submitted, dismissed and appealed. The appeal was withdrawn in July 2010. One of the grounds of the review was that the council and the NRR were swayed by commercial considerations in granting planning permission and did not give enough weight to environmental matters:

> We ... object to the fact that there appears to be a very considerable cross over in activities between NRR and BaNES and would question the legitimacy and correctness of a planning authority of a local authority (in this case BaNES) being the final arbiter in a planning application from a company

(Bellway) which is closely linked (allegedly as a joint developer) with a regeneration company that appears to be indistinguishable from the local planning authority. If in doubt, we would cite, for example, the fact that NRR letters are franked by BaNES and the organisation is hosted in buildings belonging to the council.

(Radstock Action Group, letter to BaNES, 19 March 2009)

The relationship between BaNES and the NRR is not an unusual one and is a characteristic of partnership-led regeneration, where such bodies are common. However, unpicking the concerns of the NRR, there is a feeling that the 'public interest' function of planning was compromised, though remained necessary if planning was to maintain support and legitimacy. On the other hand, planning, as part of the function of local governance, was moving from being a purely regulatory function to take on a more proactive role in helping deliver change. The problem was that planning fulfilled both functions – regulatory and proactive – within the same umbrella organisation. The establishment of arm's-length bodies did not sufficiently overcome the problem. On the other side, there was frustration that what was seen as a legitimate and desirable development vehicle was being thwarted. As the Chair of the NRR put it:

> We are powerless to do anything to speed up this process. It is her [the claimant's] democratic right to take this action, but it has already delayed the project by an additional nine months with the latest court hearing still two months away. It is also costing a huge amount of public money to defend. That money would otherwise be available for further regeneration in Radstock.
>
> *(Cate le Grice-Mack, chair of NRR, in* This Is Bath*, 13 May 2010)*

This issue of different interpretations of policy and objectives was a key point of friction between local opposition and the NRR and BaNES. The Radstock Action Group, which had consistently opposed development on the site, had a very different interpretation of sustainable development to that of the NRR and BaNES. For the group, sustainable development concerned a narrow emphasis upon environmental issues in general and the energy efficiency of the proposed houses and impact upon natural habitats in particular (Radstock Action Group, 2009). It was able to justify this position in relation to its own interpretation of national and local policy guidance. BaNES and the NRR, on the other hand, took a wider perspective of sustainability, emphasising the broader benefits of development to social and economic regeneration. Two points arise from this mismatch. First, as with the King's Cross scheme, it is possible, if not inevitable, that different interpretations emerge once the full implications of development (i.e. an actual application) become apparent. At a policy level the scope for ambiguity is significant. In response to the BaNES Core Strategy, the Radstock Action Group commented: 'Much of the text is very general; whilst it often appears reasonable, it doesn't define what will actually happen once the Core Strategy is agreed, it is sufficiently ill-defined to allow multiple interpretations of most points and proposals contained' (Radstock Action Group, 2010: paragraph 3.2).

Second, guidance provides little in the way of advice on how such competing issues can and should be reconciled. The suspicion was that both BaNES and the NRR were not 'objective' or acting in the wider public interest in attempting to balance competing interpretations, as they were partners in the development. For example, the Radstock Action Group argued that the proposals were contrary to the (then) guidance in PPG15 on the need for developments to 'preserve or enhance' the character of a Conservation Area. The NRR thought otherwise. BaNES and its conservation and urban design advisers also seemed unsure on how to reconcile the benefits of a regeneration scheme against the impacts upon the Conservation Area. Specific problems arose around the mismatch between planning policy and development control within BaNES. Despite the centrality of the authority in establishing the NRR and the supportive policy context at national, regional and local levels, a range of difficulties arose in the process to secure planning permission. While it was the government's view that planning permission should flow from the policy framework, both local protestors and planners perceived a hiatus between the two that allowed, in the case of the former group, an opportunity to thwart proposals.

Any exegesis of a complex and multi-tiered policy framework can justify a range of arguments. Table 4.2 sets out the policy framework that contextualised the NRR scheme. As with the King's Cross development, what was striking was the sheer extent and complexity of advice and guidance. Nevertheless, what would seem to be as important for securing delivery of urban policy and regeneration would be criteria and guidance on when the inevitable differing interpretation of policy emerges. As one NRR board member put it: 'There comes a point when minority interests have to be balanced against the wider community interests' (personal interview).

On reflection, some NRR board members felt that there had been too much consultation and that this had both raised expectations of what could be achieved and allowed those determined to thwart the scheme time to marshal arguments and support:

> [an objector] calls the NRR consultations a sham because he does not get all he wants from our plans. NRR listened to all comments, juggled the concerns and produced plans with Bellway that got unanimous all-party support at the planning committees. Everyone made compromises.
>
> Protest groups are not even quangos, they are formed of individuals with very different interests. They do not have to be representative. They need not consider costs. They are very valuable to the decision-making process, but when, however well amended, the application passes they feel let down. The most passionate can feel betrayed by any compromise.
> *(Catherine Whybrow, NRR board member and local councillor, in* This Is Bath, *13 May 2010)*

TABLE 4.2 Selected key events in Radstock regeneration scheme

1998	NRR formed. SRB programme awarded £2.9 million. Community Planning Weekend.
1999	Submission of outline planning permission. BaNES minded to approve the application. NRR enters into landowner negotiations to purchase the site.
2001	NRR completes land acquisition.
2002	South West Regional Development Agency funding acquired for master planning exercise.
2003	Master planning work commences. BaNES provides grant funding to the company. Further community consultation undertaken through a series of working group sessions. NRR produces a draft master plan and informs the community. NRR produces a business plan.
2004	Development partner selection process undertaken.
2005	NRR appoints Bellway as development partner. Appointment of development team. Commencement of preparation of planning application submission. Commencement and completion of site and ground investigations.
2006	Completion of 2006 master plan. Public engagement of master plan; comments reviewed resulting in some amendments. Planning application submitted.
2007	Planning application supported in principle by BaNES.
2008	Outline planning permission granted following completion of a comprehensive legal agreement.
2009	Reserved matters application approved.
2010	Judicial review of decision.

Table 4.3 raises two further issues. The first is whether any scheme can ever meet the vast extent of policy requirements, particularly if, in the case of both Radstock and King's Cross, the scheme evolves over a period when policy is constantly being changed and is at times in contradiction. The classic instance of this was the increasingly outdated PPG4 (DoE, 1992), which was clearly pro-development in general and included a presumption in favour. The replacement PPS4 was not finalised until 2009 (DCLG, 2009d). In the intervening seventeen years there was a mismatch between national planning policy on economic development and a growing concern with the more balanced (and ambiguous) notion of sustainable development. The problem with locating specific schemes within

TABLE 4.3 Selected policy context for Radstock scheme

Document	Policy context
National	
PPS1 Delivering Sustainable Development	Planning should facilitate and promote sustainable and inclusive patterns of urban and rural development, protecting and enhancing the natural and historic environment, the quality and character of the countryside and existing communities, and ensuring high-quality development through good and inclusive design and the efficient use of resources.
PPG3/Consultation draft PPS3	New development should concentrate most additional housing development within urban areas, making more efficient use of land by maximising the reuse of previously developed land and the conversion of existing buildings. Planning authorities should encourage mixed and balanced communities by ensuring that new housing developments help to create a better social mix and the delivery of affordable housing following an assessment of local housing needs.
PPS6 Planning for Town Centres	Enhance the vitality and viability of existing centres through promoting their growth and by focusing development in such centres and encouraging a wide range of services in a good environment, accessible to all.
PPS9 Biodiversity and Geological Conservation	Planning, construction, development and regeneration should have minimal impacts on biodiversity and enhance it where possible.
PPG13 Transport	By shaping the pattern of development and influencing the location, scale, density, design and mix of land uses, planning can help to reduce the need to travel, reduce the length of journeys, and make it safer and easier for people to access jobs, shopping, leisure facilities and services by public transport, walking and cycling.
PPG14 Development on Unstable Land	Ensure proper safeguards for development on land which has been damaged by mining or other industrial activities.
PPG15 Planning and the Historic Environment	Planning processes should reconcile the need for economic growth with the need to protect the natural and historic environment, particularly in Conservation Areas.
PPG16 Archaeology and Planning	Appropriate management of archaeological remains is essential to ensure that they survive in good condition. In particular, care must be taken to see that archaeological remains are not needlessly or thoughtlessly destroyed.
PPS23 Planning and Pollution Control	The remediation of land affected by contamination through the granting of planning permission should secure the removal of unacceptable risk and make the site suitable for its new use.

TABLE 4.3 *Continued*

Document	Policy context
PPG24 Planning and Noise	Need to minimise the adverse impact of noise without placing unreasonable restrictions on development.
PPG25 Development and Flood Risk	Need to assess and manage flood risk.
Regional	
Regional Planning Guidance for the South-West (RPG10)	Promotes more sustainable patterns of development through concentrating expansion in the principal urban areas and other towns, maximising the opportunities for new housing within urban areas, especially on previously developed land, and providing a range of types of housing and tenure options that reflect local conditions.
Draft RSS for the South-West	The scale and mix of development should increase self-containment of a place, develop its function as a service centre, especially in the terms of employment and service accessibility, and secure targeted development which can address regeneration needs.
Regional Economic Strategy for the South-West	Encourages community regeneration programmes to tackle long-term structural weaknesses, including levels of affordable housing, and supports provision of economic facilities in small towns and villages.
West of England Sub-Regional Housing Strategy	Need to increase the provision of new dwellings, particularly in the social rented sector.
West of England Housing Need and Affordability Model	Need to increase housing supply to improve housing affordability.
Sub-regional	
Joint Replacement Structure Plan	Promotes local community involvement in schemes to develop and regenerate urban communities, using sustainable urban locations which make the best possible use of existing resources and infrastructure by encouraging the reuse of land and buildings, including decontamination of land where appropriate. It also promotes the provision of residential development involving a mixture of types, in locations with convenient access to employment, services, facilities and open space by means other than the car, and improving the operation of public transport.
Community Strategy	Sets out five shared ambitions as to how public, private and voluntary organisations will work together, including actions identified to make improvements.

Continued overleaf

TABLE 4.3 *Continued*

Document	Policy context
Local	
Bath and North-East Somerset Local Plan (revised deposit draft)	Promotes sustainable development requirements which permit development that minimises the need to travel; meets social needs; helps employment; is of high quality design; conserves natural resources; and uses brownfield land and minimises pollution. Seeks a significant proportion (30% in the original draft version) of dwellings to be affordable. Restricts development which adversely affects sites of nature conservation value, unless material factors are sufficient to override the local biological and community/amenity value of the site, and compensatory provision of at least equal value is made. Allocates the site for development and seeks a comprehensive mixed scheme, including leisure, residential, employment and community uses and retail outlets within the town centre shopping area. In summary, development should include: • the provision of about 100 dwellings • the provision of amenity and public open space • the provision of a public transport interchange • an ecological compensation and management plan • the remediation of land contamination. • The site lies within a Conservation Area. Development will only be permitted where it preserves or enhances the character or appearance of the area, in terms of size, scale, form, massing, position, suitability of external materials, design and detailing.
Joint Local Transport Plan	Seeks an efficient, sustainable, integrated and safer transport system by making the best use of existing infrastructure and cost-effective maintenance programmes.
BaNES Affordable Housing Supplementary Planning Guidance (SPG)	The SPG sets guidance thresholds for the provision of affordable housing and negotiates the provision of 30 per cent of the total dwellings proposed for affordable housing.
BaNES Radstock Regeneration Principles	Regeneration principles for Radstock, which include the aim to create an attractive and successful town centre. The paper sets out the challenges for Radstock and the need to establish a sustainable hub with a clear role within the wider area. To achieve this will require a physical form that enables the town to develop and adapt to meet current and future circumstances.

Note: This was the policy context at the time of the reserved matters application. It subsequently changed.

complex policy contexts was recognised as a problem. The interim report from Kate Barker argued in 2006 that:

> Since 1990, regulation has continued to grow, with major new policies in areas such as transport and out-of-town centre development, as the planning system is often used as the means to give practical effect to a wide range of policy objectives ... Some Planning Policy Statements are accompanied by lengthy best-practice guidance notes, and there are still thousands of pages of national policy and guidance, including circulars and good practice guides, though where detail removes ambiguity this can be desirable. There can also be uncertainty at local level about the status of these different documents. It has taken over two years to update just nine of the 25 national policy guidance notes – completing the task could take another five years.
> *(Barker, 2006b: paragraphs 3.30–3.31)*

The Planning White Paper (DCLG, 2007a) committed the government to reduce the extent of national planning guidance and the Killian–Pretty review (2008) made recommendations, accepted by government, that the extent of supporting information requirements needed to be more proportionate (DCLG, 2010d).

The second point to arise follows on from the concern around supporting information. The 'onus of proof' is upon the applicant to demonstrate that a scheme meets disparate and often ambiguous objectives, which the local planning authority assesses through supporting information requirements. Such information requirements are costly and are largely process-driven exercises that involve little, if any, public involvement (Allmendinger, 2010). However, a more fundamental issue is the extent to which they represent a form of 'contracted-out' planning. The evidence required to demonstrate conformity and impact is mostly handed over to consultants and specialists, shifting the debates from the political to the technical arena.

Conclusions

At one level King's Cross and Radstock are two very different schemes. However, both highlight the problems with urban policy under Labour. First, despite the alignment of planning policy objectives at different scales, the opportunities for local involvement, and the approach, particularly in the NRR case, of a partnership between the public and private sectors, both schemes experienced small groups of objectors who were willing to use any mechanism to thwart redevelopment. The basis of the objections was articulated in the language of planning and the challenges depended upon discourses that were as vague as those around the 'urban renaissance'. Thus, in both cases those opposed to development could point to competing notions of sustainable development and regeneration as a basis for attacking the schemes. In the NRR case the local authority concluded that the development was in accordance with the policy framework, while the Radstock

Action Group argued that: 'The disregard for all the strategic planning contained in the Local Plan and the Regional Spatial Strategy raises the question of why these documents exist and what their function is if they can be ignored in local particular cases' (Radstock Action Group, 2009: paragraph 4).

This mismatch in the policy framework can be explained partly by differences in interpretation of what is meant by notions such as sustainable development. However, the rupture goes much deeper and reflects issues around the role of the state in the provision of services. As one local resident put it in relation to the King's Cross scheme:

> The people living and working around King's Cross/St. Pancras want that land to be used for ordinary people in the form of low cost and good quality housing and amenities, like nurseries, health centres and facilities for youth, plenty of public open space and much lower heights of buildings.
> (Camden New Journal, 18 December 2008)

This leads to the second conclusion that can be drawn from both schemes, which concerns the unsuitability of the current planning system as a vehicle to deal with such conflicts. The delivery of urban renaissance or regeneration becomes caught up in an ideological battle, which Labour considered could be tackled through partnership- and consensus-based approaches. Rather than facilitating development, the use of the mainstream development control process could actually undermine delivery.

It is worth noting that Argent anticipated securing planning permission by September 2002 following their appointment as developer two years earlier (Argent, 2001). The first application was finally submitted in May 2004 and the secretary of state's decision on the Triangle Site was issued in July 2008 – a development control process of over four years and a planning process of nearly eight. Having lost the argument at the development plan stage, the development control, appeal and judicial review stages provided a second, third and fourth bite of cherry for those intent on thwarting the development. In the Radstock case the development control period took four years. The public costs to date of the Radstock scheme, including staff time, application costs, supporting information requirements, publicity, etc., are in the region of £3 million. The costs of the public inquiry and judicial review for the King's Cross scheme were in the region of £4 million. The report on the King's Cross scheme to the Camden Development Control Sub-Committee in March 2007 ran to over 900 pages and deliberations took two days.

The time and cost of delivering regeneration and urban policy through the mainstream planning system are beginning to have consequences. The partnership-led approach of regeneration, which began under the previous Major administration, emphasises that 'Effective regeneration cannot happen without support from – and full engagement with – the people and communities involved. Local Government is uniquely placed to ensure that the needs of local residents and businesses drive regeneration' (DCLG, 2009a: paragraph 7). While it is difficult to disagree with these sentiments, they invite expectations of consensus that are unrealistic: 'Consultation does not equal consensus. Everyone's view is not equally valid. Front loading does not save time and avoid conflict down the line' (personal interview, Argent, 2009)

The role given to planning in delivering urban policy was one of growth management. This contrasted with the erstwhile understanding that recognised the irreconcilable nature of attitudes towards change and development. Consensus was and is the exception, though urban policy under Labour tried to mask differences. Where or, more likely, when consensus is not reached, then legitimacy is eroded: 'one thing that has frustrated me is that we haven't been able to manage local expectations from the start' (personal interview, NRR board member). Another outcome is the restless search for compromise to achieve consensus which privileges some local views over others. Even where such perspectives are not concerned with using the planning process to stop development, there can be significant implications for any scheme, particularly financial viability: 'Developers don't trust the public sector to make judgements – they cave in to the lowest common denominator' (personal interview, Argent). As a result regeneration through planning during the Labour era became sclerotic. The process-driven focus upon consensus and partnership had no 'end point' or indication of diminishing returns. Nor did it account for those fundamentally opposed to any development. Despite the national desire to effect change through the creation of new objectives and rule sets, such desire was insufficient, particularly when it came up against existing institutional arrangements.

Labour's urban renaissance agenda was a worthy and genuine attempt to address a wide range of social, environmental and economic issues by making cities 'liveable' and attractive places. There were numerous successes. However, the aspirations were not followed through with mechanisms to allow the easy translation of national policy into delivery on the ground. It was grafted onto an outdated and unsuitable vehicle – the post-1947 planning system – and assumed that partnership and consensus at the local level would naturally follow.

5
SPATIAL PLANNING

Introduction

To some it seemed that the Treasury-inspired, regulatory view of planning which dominated the Thatcher and Major eras had seamlessly continued under New Labour. The Confederation of British Industry and the Treasury, armed with its McKinsey report on the negative impact of planning controls on UK PLC, appeared to be setting the agenda for planning through the Planning Green Paper (DTLR, 2001a). The ministers responsible for planning at the time, Stephen Byers and Lord Falconer, talked up the prospect of reform:

> The present planning system is too complicated, too slow and engages insufficiently with local communities. We need to make it more efficient and more accessible so that it better serves everybody with an interest in the growth and development of their community, whether individuals, businesses or representative organisations.
> *(Lord Falconer,* Hansard, *12 December 2001: Column WA219)*

Yet, within four years of the Green Paper, a fresh dominant attitude within government had emerged which was supportive of a new form of and role for planning. Spatial planning, as it was termed, would work towards the creation of sustainable communities, and planners would be located at the hub of governance charged not with planning but with place-making (ODPM, 2005a). How did this volte-face come about?

Spatial planning was an enterprise of great pith and moment. It was advanced by some in the planning profession and local government as an approach for 'new times' that addressed the problems and opportunities of sectoral and spatial complexity and the kind of delivery issues being experienced at King's Cross and Radstock. These spatial planning advocates had clear normative preferences around progressive styles of governance and found succour from the theoretical Zeitgeist within parts of academia that had embraced post-positivist theory and talked of 'spaces of flows', networked governance, collaborative practices and scalar complexity. Such an analysis was packaged and presented so as to echo the sentiments of New Labour and its concerns with integration, collaboration and, most of all, the delivery of growth.

Was spatial planning a genuinely new paradigm or a temporary experiment endorsed and facilitated by parts of New Labour because it chimed with its vision for local governance and allowed 'soft', flexible and relatively cheap responses to politically sensitive issues? The answer to this lies in what is meant by spatial planning. There is a strong argument that, as a practice and state of mind, spatial planning was a label for what planners did and had been doing for some time. At another level, there is also an argument that the (misleading) link between it and the 2004 Act undermined spatial planning, particularly following the failure of local planning authorities to make progress on the new system of development planning. The result was that the failure of the post-2004 approach tainted spatial planning and led to a backlash against both spatial planning and the 2004 Act. (Allmendinger and Haughton, 2009b, forthcoming, 2010). The rush to celebrate and promote the 'renaissance' of planning has led to some selective rewriting of history, particularly the claim that the 2004 Planning and Compulsory Purchase Act introduced the notion of spatial planning. What has also been subsequently overlooked is that, even at the time, there was by no means universal support for the changes: the Town and Country Planning Association, for example, felt that spatial planning either did not address or failed to understand the fundamental problems of planning at the time (Inch, 2009).

While embraced by some in Labour (though not initially), the movement for spatial planning was led by a loose coalition of interests within local government and the profession as a 'solution' to at least one problem. The problem was portrayed as being around the need to work within fragmented sectoral and governance frameworks in order to 'create a less formal, more creative, more integrative approach to planning' (Upton, 2006: 112). In reality, the 'problem' was actually more to do with finding a coherence and direction for planning in the face of searching, Treasury-led inquiries and the possibility that planning as a public-sector activity was in danger of being thoroughly 'modernised' – i.e. deregulated. There is a still a feeling that, despite the post-2004 changes, planning remains largely within its original 'command and control' orientation (Zetter, 2009). Ironically, the promotion of spatial planning might expose the unsuitability of current regulatory mechanisms to manage change and growth and could have hastened more fundamental reforms, which are looking more, not less, likely under the coalition government.

This chapter explores the development and differing notions of what is meant by spatial planning before turning to examine two examples of it in practice. Both cases highlight the marginalisation of what is widely regarded as being 'spatial planning' as a vehicle for growth management. The success of growth management has more to do with arrangements for governance between a variety of mainly public bodies, allied with a coherent and agreed vision and backed up with infrastructure. The role of planning, perceived as being the 'glue' in this process, is actually minimal. The cases also illustrate what those involved in planning regard as being the Achilles heel of spatial planning – development control. The 2004 Act largely ignored this element of planning, assuming that development control would be a continuum of development strategies. Notwithstanding attempts to rebrand development control as 'development management' after 2008, in a largely futile attempt to make up for the misunderstanding of how the planning system in the UK operates, securing planning permission, even for those schemes that are part of the development plan in the spatial planning era, remained an uncertain, costly and lengthy process.

What is spatial planning?

There have been several studies published that link the origins of spatial planning in the UK to continental European experiences and approaches (Shaw and Lord, 2007; Baker *et al.*, 2007; Nadin, 2007). Such linkages are largely *post hoc* justifications, and the concerns of strategic (both long term and large scale), holistic (environmental, social and economic) and integrative planning could be argued to go back to early pioneers of planning thought such as Patrick Geddes. As one observer from elsewhere in Europe has pointedly put it:

> I rather tend to believe that the introduction of spatial planning [in the UK] ... is caused by the ambition to reinvent, to rejuvenate planning in Britain under a new label, which is not loaded with all the negative experience of post-war planning in the country. As often happens in the planning community, when a new buzzword is emerging, it inspires the imagination of planners, who subsequently load this concept with all their desires and use it to address planning problems, which previously could not be addressed.
>
> *(Kunzmann, 2009: 142)*

Rather than providing a distinctive approach to planning, the term seems, instead, to have developed as a neutral way of describing the multitude of different approaches to land-use control across Europe (Faludi and Waterhout, 2002). However, the link to continental practices does provide a useful back-story, justification and borrowed credibility (Allmendinger and Haughton, 2009b; Haughton and Allmendinger, forthcoming, 2010). It has also meant that the definition of 'spatial planning' is usefully vague, allowing a wide range of stakeholders to sign up to the concept. The characteristics of spatial planning can be taken to include:

- re-engagement with long-term planning as a means for guiding spatial development. In this sense, spatial planning reasserts the value of thinking spatially and for the future;

- re-engagement with a variety of other sectors and professions;
- a new legitimacy for planning and the profession via the discourses of sustainable development, incorporating aspirations to make future development more environmentally sensitive while also supporting parallel goals to improve the social well-being of people and economic competitiveness of places;
- an attempt to redefine the relationship between planners, a wide variety of 'stakeholders' and the general public. The new relationship takes a less technocentric and elitist approach, which values diverse knowledges and aspirations, facilitating a broad dialogic process that involves creating systems to allow the input of planning expertise without asserting its primacy over other forms of knowledge and value systems (Haughton *et al.*, 2010).

This vagueness can be criticised for seeking to be all things to all people (McGregor, 2004) and for trying to create a break with a mythical past through a 'planning renaissance'. A strong theme running through definitions of spatial planning is the need to define it with reference to what it is not. Part of the folklore of planning in the UK in recent years concerns its neoliberal reorientation during the 1980s, which I discussed in Chapter 1. While there were undoubtedly changes to planning driving this era, their extent and impact were far less than feared. Nevertheless, the portrayal of spatial planning in the 1980s as the antithesis of regulatory planning serves as a useful device to distinguish it from planning in the past and to garner support, but without providing clarity. This ambiguity of purpose is echoed in one of the evaluations of spatial planning, which highlighted that 'it is apparent that amongst key participants including planners, councillors, senior local authority and other public sector managers that there is little common understanding what this means in practice' (RTPI, 2007: 1). The confusion of purpose led to the emergence of a range of interpretations.

Planning as the spatial Third Way   There was a strong though loose lobby for spatial planning within academia, local government and the planning profession bound up with a normative purpose for planning as an 'active force for change' (Newman, 2008: 1373). Inch (2009) argues that this discourse coalition used the vague notion of spatial planning successfully to counter the negative, deregulatory instincts of the Treasury. As part of the justification for this approach, the advocates of spatial planning created a false opposition, contrasting spatial planning with 'traditional' approaches where there was little or no vertical or horizontal integration between the different levels and parts of government (Albrechts, 2006a: 1158; Local Government Association, 2000). Spatial planning, on the other hand, was portrayed as progressive, proactive, democratic and deliberative (see Healey, 2006a, 2006b; Albrechts, 2006a).

As Gaffikin and Skerrett (2006) note, such critiques of 'traditional planning' were actually closely bound up with carving out a greater role for planning in countering the dominant economic and physical objectives with the social and environmental ones. Spatial planning was an attempt to create a new direction for planning based upon redistributive and progressive goals of the political left (ibid.: 161). In this perspective, it was portrayed in similar language and with equal vagueness to Third Way thinking, promising 'win–win–win' solutions.

Spatial planning as a rebranding exercise   Labour did not come to power with a position or even any clear views on planning. This vacuum was filled by the deregulatory and critical perspectives of bodies such as the CBI, which argued that planning was a supply-side constraint upon growth and competitiveness (CBI, 2001). Labour in general and the Treasury in particular were sympathetic to business views and, with little evidence to the contrary, began to form an agenda based around them. Early initiatives on planning sought a rationalisation of plans and the subsuming of development plans into Community Strategies (Allmendinger and Tewdwr-Jones, 2000). Planning was under threat. Spatial planning, therefore, was a rebranding exercise in order to counter what was rightly envisaged as a critical and possibly radical agenda that was looking at alternative models for land-use regulation. While such an understanding is more cynical, it is also more realistic in its claims about the impact of Labour upon planning. The 2004 Act and subsequent policy changes were built upon the foundations of a system and a culture of planning that had remained largely untouched in many important respects since the end of the Second World War. The gap between expectations of spatial planning and the reality can therefore be explained as analogous to changing the marketing of a product rather than the product itself. Rebranding planning as spatial planning was made possible by the alignment of the 2004 Planning and Compulsory Purchase Act, which introduced the new system of development planning and the new, more sympathetic perspective of the then deputy prime minister, who had taken overall responsibility for planning. Grafting the 'new' approach onto the post-2004 approach provided the opportunity to present planning as a way of achieving a range of policy objectives.

New planning for new times   Echoing the pragmatism of many policy responses under New Labour, spatial planning can be seen not as a normative response to new times but as a natural and pragmatic reaction to the day-to-day realities of institutional and sectoral fragmentation. The devolutionary movements within Labour were accompanied by a top-down and detailed performance target culture. Nevertheless, New Localism (Lyons, 2006) as a broad movement was backed up and facilitated by a range of policy tools and mechanisms which called for a move away from the 'one size fits all' model of local services to more bespoke approaches built around local needs and uniqueness. While local government reorganisation continued under Labour, it became increasingly clear that a related issue concerned the changing nature of local government and the need to look beyond local government to local governance (Stoker, 2004).

This rethinking of the nature of local government echoed many of the concerns of relational geography (see, for example, Amin and Thrift, 1995; Massey, 2005). Such perspectives argue that fragmentation and networked relations are essentially unstable and changing – there is no single, ideal functional administrative or planning area. Instead, there are multiple such areas, and the fixing of any one through, say, local authority boundaries, privileges some interests and relations over others.

According to some, functional planning areas prioritise economic interests over the social and the environmental (Healey, 2009). Under this view of the world, spatial planning becomes a way of replacing formal jurisdictions with bespoke, locally derived processes that take account of and help deliver 'place-making'. Spatial planning is portrayed as an alternative to the restless search for functional planning arrangements and as a neutral approach to tackling the increasingly globalised networks of relations (some close, some far away) and the dominance of the Treasury-led economic discourse that drives policy.

Spatial planning as neo-traditionalism   Spatial planning is nothing new but a re-emphasis upon linking plans to implementation (Goodstadt, 2009). Rather than being a renaissance in planning, it is a case of 'back to the future'. The much vaunted sectoral coordination of spatial planning, for example, is part of an established belief in the need for policy synchronisation. While early moves towards more corporate forms of local governance in the 1970s were driven as much by financial restraint as by ideology, the effect was the same. The creation of government offices in 1994 and initiatives such as the Single Regeneration Budget introduced in 1993 also privileged coordinated approaches. If the theme of sectoral coordination is hardly new, then the notion of networked spatial governance is also recognisable to the planner who prepared structure and local plans under the influence of 'corporate planning' (Friend *et al.*, 1974; McKay and Cox, 1979; Bruton and Nicolson, 1987).

So how does spatial planning differ? As Peel and Lloyd (2007) argue, the 'new' spatial planning is largely built on the legislative, policy and institutional infrastructure of more 'traditional' planning. Spatial planning is a welcome return to how planning was and should be undertaken. Prior (2005) goes further and maintains that spatial planning has a similar function to planning in resolving the underlying contradictions in capitalist economies (see Taylor, 2009, and Inch, 2009, for a similar analyses).

Neo-traditional perspectives on spatial planning tend to point to the role of planning in addressing irreconcilable tensions within the mode of production: '... reforms have either served to suppress (through the reforms of the 1980s) or hold in balance (through the current reforms) the capitalist–democracy contradiction but, in both cases, in ways conducive to the reproduction of capitalism' (Prior, 2005: 481). These discourses around the purpose and characteristics of spatial planning are not distinct or mutually exclusive and sometimes draw upon elements of each other to reinforce the general movement towards spatial understandings. However, they are built upon different reactions and perceptions of change. The Third Way spatial planning school was an attempt to capture the agenda for planning and promote a particular world-view, drawing upon the analysis and language of modernisation propped up by a back-story of spatial theory and continental

European practice. On the other hand, those who point to spatial planning as a rebranding exercise or highlight the lack of substantive change may not share the same normative purpose: they too are engaged in a battle over the nature of planning.

There are clearly problems with identifying what spatial planning is, and undoubtedly this ambiguity is useful in some ways. The notion of spatial planning has led to a range of approaches, echoing Stoker's (2006) idea that New Labour deliberately introduced experimental change as part of its modernisation agenda. Yet one cannot help thinking that the ambiguity around the notion of spatial planning on the part of government had less to do with experimentation and more to do with placating the Treasury-inspired view that planning was an outdated constraint upon development and competitiveness. This ambiguity is not helped by the confusion between the 2004 Planning and Compulsory Purchase Act and the notion of spatial planning. According to Baroness Andrews, former junior minister for planning, the 2004 Planning and Compulsory Purchase Act

> establishes a process it [the government] calls 'Spatial Planning' ... It facilitates and promotes sustainable and inclusive patterns of urban and rural development. Rather than operating through a narrow technical perspective, spatial planning should actively involve all members of society because everyone has a stake in the places in which they live, work and play.
> *(Quoted in RTPI, 2007: 6)*

Yet the Planning Green Paper of 2001 made virtually no mention of spatial planning and even managed to dismiss it as a rather technical approach: 'In planning jargon, it [the Local Development Framework] would be much more of a "spatial" strategy' (DTLR, 2001a: paragraph 4.12). The then secretary of state, Stephen Byers, described the proposals in the Green Paper as the biggest shake-up of planning for more than fifty years. The current system was, he concluded, 'slow, ponderous and uncertain' (DTLR, 2001b: 1). The House of Commons Select Committee that examined the Green Paper was similarly unconcerned with spatial planning *per se* as a concept. In their memorandum to the House of Commons Select Committee, the Royal Town Planning Institute (RTPI) broadly supported the new system of development planning, though it argued that a functional sub-regional tier of spatial strategies to match the Regional Spatial Strategies (RSSs) would also be required. Again, however, no mention was made of spatial planning. The notion of spatial planning as debated around the 2002 Planning and Compulsory Purchase Bill was specifically related to the regional level.

The announcements by the deputy prime minister in 2002 on how the government intended to take forward the Green Paper did not mention spatial planning, though they did backtrack on some of the more business-orientated aspects of the changes. Given the above, it is clear that spatial planning as a notion did not underpin the changes introduced or drive government thinking. Spatial planning and the 2004 Act emerged coincidently, though actually bore little relation to each other. Indeed, spatial planning as an approach owes much more to changes in policy guidance and the creation of a new doctrine for planning than the new system of development planning. The erstwhile regime of development planning could easily have been orientated towards a spatial approach without the need to complicate change further by introducing a new system.

This is not to say that the post-2004 approach to development plans was not significantly different: the emphasis upon a suite of documents, the relative ease of updating, programme management and the 'front loading' of public involvement were departures – in theory if not in reality. However, key components such as the test of soundness used by the Planning Inspectorate were introduced subsequently and did not relate to the original intentions of the new approach, which had been developed with different objectives in mind. Equally, the link to Community Strategies (later Sustainable Community Strategies) built on the notion of spatial planning, not the new system of development planning. The report into planning by the Royal Commission on Environmental Pollution, which was initiated as a response to the Planning Green Paper, advanced the idea of Integrated Spatial Strategies to which development plans should conform. These spatial plans would be founded on functional planning areas or sub-regions and help to deliver sustainable development. In reality, Integrated Spatial Strategies looked remarkably similar to the Community Strategies that had been introduced in the Local Government Act 2000. The latter were to provide the longer-term strategic vision for an area, and development plans – specifically Local Development Frameworks (LDFs) – were to set out, in the words of the Select Committee on the Planning Green Paper, the 'land-use' aspects of proposals in Community Strategies (HoC TLGR, 2002: paragraph 38) – a much more perfunctory role than spatial planning.

There was widespread unease at the proposed changes to development planning at the time. The feeling was, at best, that the proposals would not achieve the government's objectives and, at worst, that they would make the system slower and more complex. Spatial planning as a notion did not figure highly in the legislative changes, which were aimed more at speed of preparation and simplicity. The new system was not the vehicle that would have been designed to facilitate a more spatial approach. Such a vehicle, as the RTPI and others pointed out, would involve more functional, sub-regional development planning. The separation of Regional Spatial Strategies from Regional Economic Strategies also did not bode well for truly spatial planning, with each being prepared by a different body under different processes and timescales.

Provision was made in the 2004 Act, however, to facilitate plan-making across administrative boundaries, through either joint plans or, as we shall see below, bespoke plans combined with development control functions. What remains unclear is how and why there was a rather sudden departure from the largely unwelcome system introduced in the 2004 Act and the subsequent acceleration towards spatial planning. From 2004 onwards, spatial planning as an approach began to be grafted on to the new development plan system, regardless of the objectives of the legislative reform. Government advice in Planning Policy Statement 11 (ODPM, 2004c) and PPS12 (ODPM, 2004d) was relatively restrained in the promotion of RSSs and LDFs as spatial plans, emphasising the need to 'integrate policies for the development and use of land with other policies and programmes which influence the nature of places and how they function' (ibid.: paragraph 1.8). The need to adopt a 'spatial planning approach' to preparing LDFs was left vague, though reference was made to the Planning Officers' publication on the subject (Planning Officers' Society, 2005). Mention was also made of the link between Community Strategies and LDFs (ODPM, 2004d: paragraph 1.10).

It became increasingly clear that spatial planning, far from producing quicker, simpler plans, would require much greater attention and input from planners and developers in ways that remained largely unclear:

> Conventional approaches to making development plans will not result in true spatial plan policies. Instead, planners will have to adopt new approaches and techniques for the job. This will involve you in abandoning many of your long-established ways of working.
> *(Planning Officers' Society, 2005: paragraph 4.1)*

The result was understandable confusion. The objectives of the 2004 Act around simplicity and speed had been reversed through policy changes, backed up by a new, strengthened policing role for the Planning Inspectorate through its soundness tests (PINS, 2005). The confusion was not restricted to local authorities. Advice on the new approach provided by the government offices to local authorities preparing LDFs differed to that from the Planning Inspectorate, as the finding of the Stafford and Lichfield core strategies as 'unsound' in 2006 demonstrated.

Experiences of spatial planning

It is perhaps no surprise, therefore, that experiences of spatial planning were, at best, mixed. Even advocates concede there was little in the way of progress:

> Despite much rhetoric about 'integrated' policy agendas, so far there has been little sign of the development of locally specific place development strategies which bring into conjunction the social, environmental and economic dimensions of the 'place development' of 'city regions'.
> *(Healey, 2009: 839)*

Despite the widespread dissatisfaction with the notion of spatial planning, particularly in practice, there are problems in its evaluation. First, the vagueness of what it is provides few criteria against which to measure spatial planning. Second, the link in the minds of many between spatial planning and the post-2004 development plan system means that assessments tend to conflate the two. The government-sponsored evaluation united spatial planning with the post-2004 system, focusing mainly upon the latter in its conclusions (DCLG, 2008b), and the conflation was also evident in the RTPI-sponsored study (RTPI, 2007). Overall, the government evaluation struggled to find evidence that practice was achieving the objectives of the new approach, highlighting the delays in moving to the new system, the lack of 'buy-in' from practitioners and little evidence of improved consultation. There was some emerging evidence of joint working and improvements in the evidence base (DCLG, 2008b), though, as the RTPI study pointed out, there was little in the way of linkages between development plans and Community Strategies (RTPI, 2007: 27). Recent revisionist evaluations point to the weak evidence supporting the claims that spatial planning is becoming the dominant style of planning (Newman, 2008).

In addition to these mainstream reviews there has been considerable discussion of the shortcomings of the new approach to development planning. For example, Watson and Crook (2009) answer their own question – How could the civil servants have got it so wrong? – by pointing to the lack of resources within local planning authorities required to move towards its adoption, the cautious approach within authorities following rejection of some Core Strategies by the Planning Inspectorate, and the vagueness of the new regulations. Subsequent reform on the 'test of soundness' and the changes in procedure introduced in the 2007 Planning White Paper (DCLG, 2007a) underline the problems with the roll-out of spatial planning and the post-2004 system of development planning.

One issue is that it is difficult to separate spatial planning from the post-2004 system of development planning, particularly when they are taken by many to be the same thing. Where positive aspects of planning practice are highlighted, it is not clear whether this is a result of the move towards spatial planning, a continuation of ongoing good practice heralded by the new development plan approach, or simply a unique configuration of factors, including the will of local actors. Similarly, where criticisms are made that a particular place does not meet the aspirations of spatial planning, it is not clear what those aspirations are (because of the vagueness of the terms and the confusing and conflicting advice and guidance provided) or whether, given the spirit of new localism, that locality has decided to place a different emphasis or different priorities upon planning. The latter is particularly pertinent given the tension between, on the one hand, a more spatial approach to development planning and, on the other, a more target-driven approach to development control, with the latter backed up by performance targets, monitoring and financial inducements. It is also clear that few organisations actually signed up to the notion of spatial planning.

In this next section I outline two examples of 'spatial planning' against the backdrop of the new system of development plans and try to unpick the nature of planning style from the legislative requirements of development planning. In the first case, sub-regional planning in and immediately around Cambridge provides an example of how an emphasis upon delivery of a growth agenda across administrative boundaries has been created despite rather than because of the new system of development planning. Here a new legal entity for plan-making and decision-taking based around a functional planning area shows how spatial planning in this instance is actually an attempt to work across artificial administrative boundaries when a more sensible, effective, though costly alternative would be to reorganise those boundaries. It also highlights how both the role of the government office and financial incentives were significant in creating this solution. Nevertheless, the result is a complex set of arrangements that separate the visionary and implementation element of planning from the more perfunctory and contested day-to-day practices.

The second case involves spatial planning in the Thames Gateway, which is a pan-regional growth area with widely recognised institutional complexity. Here, spatial planning has become much more about creating a flexible and delivery-focused approach, much in line with the original intentions of the 2004 Act. However, planners and others there are creating shadow strategies and methods on a variety of scales outside of the formal planning mechanisms in order to avoid the inevitable delays and complexity of the new spatial approach. Significantly, what links both (very different) areas is the separation of planning visioning (or place-making) from the statutory functions of planning.

Spatial planning and the growth agenda

Cambridgeshire Horizons is widely regarded as an exemplar of a sub-regional delivery body. A former junior planning minister, Baroness Andrews, commented in a speech delivered to the Cambridgeshire Horizons conference on 8 June 2007 that 'Horizons can be held up as an outstanding example of what growth area delivery vehicles should be aiming for.' While it has no statutory planning functions, which remain vested in the local authorities, it provides a functional spatial planning body that coordinates development activity to achieve implementation of local, regional and national objectives. Established in 2004 with funding from, among others, the East of England Development Agency and (the then) English Partnerships, its role is to help deliver major growth in the county, working with partner local authorities and other actors.

The growth agenda for Cambridgeshire is significant and includes a target of 73,300 new homes and 50,000 new jobs by 2021. This level of growth will require around £4 billion of infrastructure. The proportion of growth in the Cambridge sub-region provides particular challenges, where the target is to deliver around 19,000 homes by 2016 within or on the edge of Cambridge and a further 6,000 in the proposed new town of Northstowe 5 miles north-west of the city (Map 5.1). Cambridge city had a housing stock of around 46,000 units in 2005–6 and an average house price of 7.75 times average income in 2008. Problems around housing cost and availability arose largely from the success and problems of accommodating the 'Cambridge phenomenon' of hi-tech, university-related spin-offs. The national importance of such enterprise to the economy was recognised in the Barker review (Barker, 2006a) and the report by Lord Sainsbury, as minister for

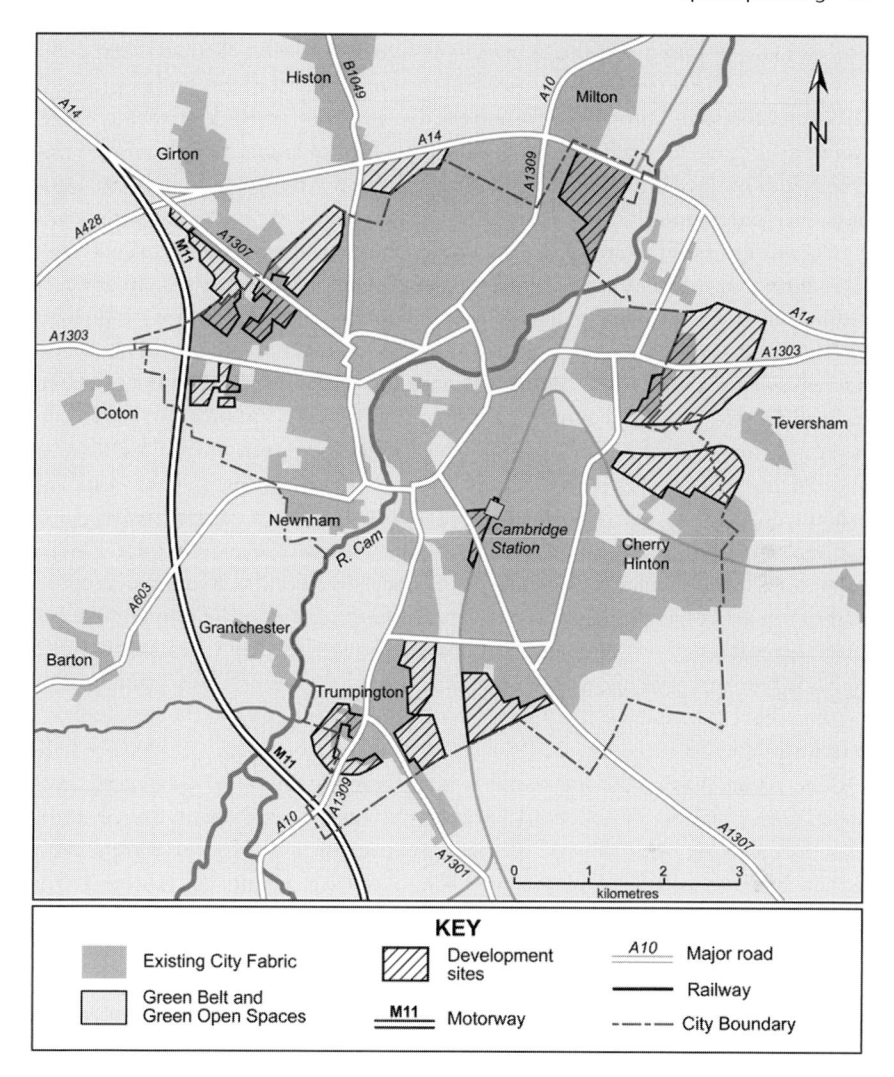

MAP 5.1 Cambridge sub-region development proposals and administrative boundaries.

science, on biotechnology clusters (DTI, 1999). Criticism from the scientific community, typified by the five years it took the Wellcome Trust to secure planning permission for a facility to undertake work on the human genome project, led to a coalition of local interests to lobby a receptive government on the need to reverse the anti-growth planning framework in the late 1990s. Until then the planning regime had been underpinned by notions of growth restraint in the city through a green belt drawn tightly around the built-up area and limited dispersal to the surrounding towns and villages. In addition to the Wellcome Trust proposal, a key moment came with the publication of an article in the *Cambridge Evening News* by Peter Dawe, founder of the internet company Pipex, who claimed that the city was 'full up' (While et al., 2004). Around the same time the Cambridge Network (1996), Cambridge Futures (1997) and the Greater Cambridge Partnership (1998) were established. All were multi-agency public–private bodies aimed at supporting the growth of the city on the back of the success of the university and the coalition of hi-tech industries.

Cambridge Futures, in particular, was established as a group which included local authority officers, academics, business leaders and politicians in order to rethink the future growth strategy of the city on the back of pressures for growth. According to planners in the city, Cambridge Futures was able to 'think the unthinkable' and challenge existing orthodoxies around growth restraint. Planners in both the city and the county had realised that growth restraint had led to car-based commuting, congestion and pollution, high price inflation and problems of affordability. They also realised that demand management policies had not restrained ongoing and future economic growth, which was increasingly of national importance. According to one involved at the time, Cambridge Futures came up with ideas that were too politically sensitive for the council to propose. Part of this process included an analysis of different options for the future of the city based around seven scenarios, from minimum growth, to new town, to densification. While not making any recommendations, the report was influential in helping shape future debates around growth and looking beyond the development restraint strategy.

The actual shift in the spatial policy for Cambridge and the sub-region came about from a number of directions and from different scales. A significant backdrop was the growing awareness of the need to plan in ways that promoted more sustainable forms of development. *This Common Inheritance* (DoE, 1990), *Sustainable Development: The UK Strategy* (UK Government, 1994), *A New Deal for Transport* (UK Government, 1998) and *A Better Quality of Life: A Strategy for Sustainable Development in the United Kingdom* (UK Government, 1999) provided the backdrop and began to show the need for higher-density, integrated and mixed-use developments.

More sustainable forms of development were not, however, the only narrative underpinning growth. Labour had come to power in 1997 committed to shedding its reputation for being anti-market and anti-entrepreneur. While embracing the notion of sustainable development, the Department for Trade and Industry published the Sainsbury report on clusters (DTI, 1999) and the Department of the Environment, Transport and Regions published *Planning for Clusters* (DETR, 2000c), while the Treasury-commissioned McKinsey report (1998) highlighted the role of planning in thwarting competiveness and growth. Both strands of thinking – sustainable development and the need to compete in a global economy through developing hi-tech clusters – came together in the form of *Regional Planning Guidance for East Anglia to 2016* (RPG6), published by the Government Office for the East of England (GOEE, 2000). RPG6 represented a shift in spatial strategy, integrating transport with development and introducing a sequential approach that would focus new development firstly within the city, then on the edge of the built-up area and, if necessary, through a new settlement. Access to public transport would be an important criterion in helping shape the location of new development. However, a critical departure from the erstwhile regime and a key element of the new spatial strategy was the commitment in RPG6 to review the Cambridge green belt.

At a local level the *Cambridge Subregion Study* by consultants Colin Buchanan for the local authorities in the sub-region was published in 2001 and included the green belt review, while Roger Tym and Partners were commissioned to examine the infrastructure and governance arrangements required to implement RPG6. Various options for development and governance were explored, and both were influential in the deposit draft of the Cambridge and Peterborough Structure Plan published in 2002 (CCC, 2002) and endorsed by the Examination in Public in 2003. The Structure Plan was underpinned by the strategy of accommodating future economic growth and linking this to housing. The forecast in the plan was for a population increase between 1999 and 2016 of 21,100 within the city and 42,700 in South Cambridgeshire, with an associated housing growth of 12,500 and 20,000 units respectively. This was to be achieved by a combination of green belt land release in Cambridge and South Cambridgeshire, development within the city, and a new settlement at Northstowe.

The upshot of this growth strategy between the three main authorities (Cambridge City, South Cambridgeshire and Cambridge County Council) was the need to turn to governance and delivery. The Colin Buchanan study had come to the conclusion that the administrative boundaries and split responsibilities of the three authorities could be a barrier to delivery, and the Roger Tym study had recommended a partnership delivery vehicle. The government office was also keen to know how such an ambitious growth agenda would be managed. Two further issues around delivery needed to be addressed for the growth strategy to be achieved. First, the government published its *Sustainable Communities* plan (ODPM, 2003a), which identified the Cambridge–Stansted–London corridor as one of the four key growth areas. It provided a new space and scale of strategy-making and a coherent narrative that sought to integrate and coordinate public and private investment and spending around an explicit growth strategy:

> There has been substantial economic growth in the London–Stansted–Cambridge corridor over the last decade. This is underpinned by clusters of some of the UK's most successful businesses in biotechnology, life sciences and ICT/software; and a rapid increase in the use of Stansted airport. The issue is not whether growth will continue, but at what level and how that growth is handled.
>
> *(Ibid.: 55)*

The alignment of a national-level growth agenda with the emerging local and sub-regional agendas further showed up the need to address how the process

102 Spatial planning

would be managed and raised a number of potential models from Urban Development Corporations (UDCs), which had been introduced in Northampton and the Thames Gateway. Under the threat of a UDC as well as the prospect of government funding for transport improvements, the authorities involved with the support of the East of England Development Agency and English Partnerships agreed in 2004 to the establishment of the partnership-based delivery body Cambridgeshire Horizons. Horizons covers the whole of Cambridgeshire, but as far as the growth in the Cambridge sub-region is concerned it is a vertical link between the two tiers of local government and a horizontal link between the contiguous local authorities of Cambridge City and South Cambridgeshire. In effect, it addresses the problem of growth across and between administrative boundaries.

The second issue concerns the gap between delivery and planning control. Horizons has no statutory powers, so planning functions are still undertaken by the relevant local authorities. While the post-2004 development planning system diverted attention and resources away from existing strategies and their implementation, the Act did contain provisions for joint development planning (Sections 28 and 29). This was in addition to the existing powers for joint development control under the 1972 Local Government Act (Section 101). Joint plan-making and development control functions now exist for a number of areas identified for development on the city fringe. Again, additional funding from government was forthcoming once the joint arrangements were in place. The upshot is a complex arrangement of joint boards and committees (Figure 5.1).

The experiences of planning for growth in the Cambridge sub-region raise a number of significant issues around spatial planning. First, much of the thinking and coordination occurred in the late 1990s, well before the explicit move towards spatial planning occurred, in the early to mid-2000s. The link in the minds of

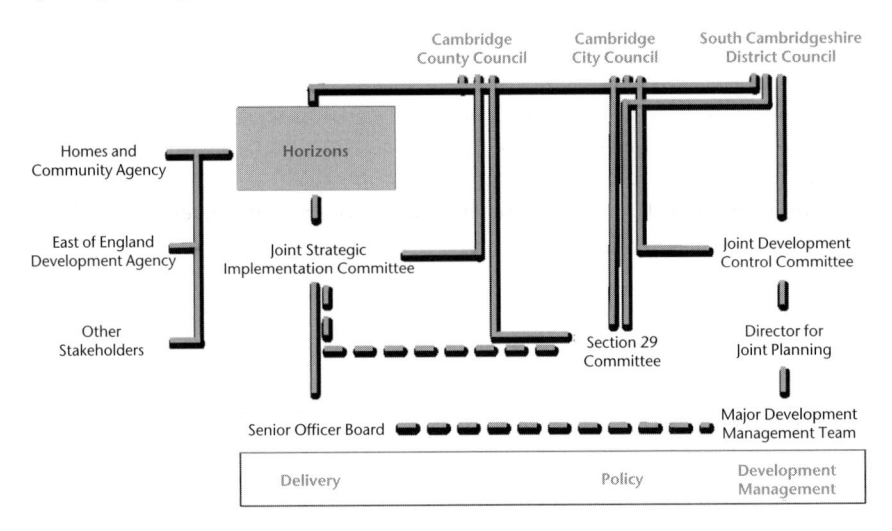

FIGURE 5.1 Organisational arrangements for growth management in Cambridge.

many between the 2004 Act and spatial planning is a case of *post hoc ergo propter hoc*. On the contrary, spatial planning – in the sense it is mostly used in the UK, in relation to a multi-spatial, coordinated and integrated process aimed at place-making – was embedded in Cambridge before the UK signed up to the European Spatial Development Perspective and embraced the notion from circa 2003 onwards. It is worth noting that it emerged during an era when national government considered planning to be largely a regulatory function. Second, the growth agenda and the consequences for planning and the wider public sector did not come solely from local authorities but from a coalition of business and other interests. Local authorities and the planning system created the legal and governance mechanisms to make real this shift in strategy. This amounted to a two-tier role for planning, with the statutory process being used to deliver upon a wider spatial planning or strategy making. In other words, the general conflation of the post-2004 changes and the notion of spatial planning confused means and ends. Spatial planning, as undertaken in Cambridge, existed apart from the legal mechanisms and processes, was embedded in a range of private and public bodies, and was coordinated through Horizons.

Second, the critical actors and statutory documents in the shift in strategy were Cambridgeshire County Council and its Structure Plan, as well as the government office with its Regional Planning Guidance. Both strategies were replaced by the post-2004 system and the role of the bodies became diluted. It is worth noting that the role of the Structure Plan was to provide a sub-regional strategy that the RSS has had to replicate. Third, the statutory planning system, both development planning and development control, along with the Local Strategic Partnership and Local Area Agreement, are marginal to the future growth of Cambridge and the sub-region. They follow and largely legitimise the strategies for growth determined elsewhere. This draws attention to the hiatus between spatial planning as a process and state of mind and the significance of the post-2004 system.

While the view that Horizons may be a 'hard solution', and a unitary authority based on a sub-regional area with the necessary powers would have been able to undertake a similar role, this misses some important aspects of the model. For Horizons itself there are advantages in not being part of a local planning authority structure. There is a distinction between the statutory planning process, including development control, and the more strategic, visionary and delivery-focused role of Horizons. For Horizons, the former can mitigate against the latter. One view in Horizons is that being responsible for the highly technical, process-driven, spatial and sectorally complex and increasingly contested statutory planning functions is a specialist task that should be separate from the focus upon strategy and delivery, as the two can conflict. In an echo of the 'soft spaces' that are evolving in the Thames Gateway, discussed further on in this chapter, the statutory processes and requirements create a process-driven arena that is slow and inflexible. A further dimension concerns community representation in partnership-led approaches. There are always a multitude of views regarding development. The statutory planning process and the planners involved have a responsibility to take into account and make decisions in the wider public interest, as highlighted in Chapter 4 around the role of the Norton Radstock Regeneration Company and that of the London boroughs in the King's Cross scheme. If there is a close alignment between the role of a local planning authority as a partner in development and delivery, then who represents alternative views? This 'critical distance', where planners and planning interface with a multitude of views and attempt to reconcile and resolve them, can place development planning and control in conflict with each other and with other strategies and bodies. The assumption underpinning spatial planning, that there would be a seamless transition from plan to permission to development, was severely misplaced: spatial plans do not reconcile divergent views, and development control provides a further opportunity to challenge and scrutinise the details of proposals. Despite attempts to bridge the gaps between plan-making and decision-taking through the introduction of 'development management' (DCLG, 2010b) and incentives in the Housing and Planning Delivery Grant, there would seem to be strong arguments for actually ensuring that the different elements of planning are not brought too closely together *à la* spatial planning and having distinct functions in different bodies.

There is a further advantage in the separation of responsibilities. The Cambridge example demonstrates that the achievement of near consensus on the need for growth and development was a necessary prerequisite to the subsequent development plans and vehicles. This 'winning the argument' was based upon what While, Jonas and Gibbs describe as a 'powerful local growth coalition and its role in reorientating the local mode of regulative development in favour of major development' (2004: 280).

Recent developments have highlighted the importance of having a strategic body able to focus upon delivery. Following the Planning White Paper (DCLG, 2007a) and other initiatives, 'mini-reviews' of regional housing targets were initiated, which led to a requirement for a further 4,000 houses to be accommodated in Cambridge and its hinterland. At the same time, the recession has affected development viability, particularly on a key city-centre site adjacent to the railway station. Finally, proposals for changing the system of planning obligations (see Chapter 7) have required coordination to ensure that strategic and cross-boundary infrastructure critical for the growth strategy as a whole is not undermined. Horizons has been the lead body in addressing each of these issues.

Spatial planning in the Thames Gateway

Like Cambridge, the Thames Gateway was identified in the *Sustainable Communities* plan (ODPM, 2003a) as being one of the government's growth areas. It extends 40 miles to the east of central London and around 20 miles north of the Thames (Map 5.2). It has around 1.45 million existing residents and 637,000 employees. Sixteen local authorities lie within, or partly within, its area, which also covers three standard government regions. In order to focus upon regeneration and development there are three sub-regional partnerships, as well as two Urban

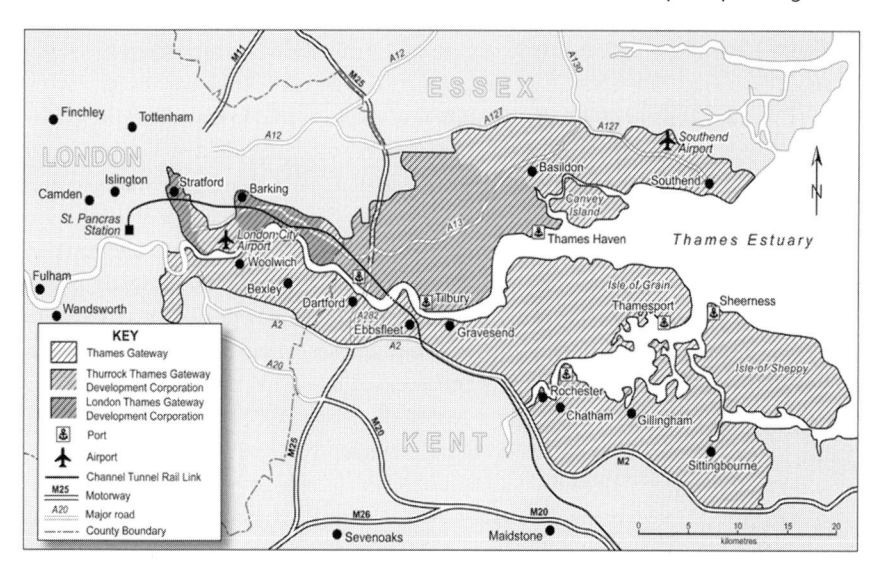

MAP 5.2 The Thames Gateway area.

Development Corporations, an Urban Regeneration Company and the Olympic Delivery Authority. In addition, there are the other local services of health, education, police, gas, electricity and a range of other agencies charged with environmental and flood protection, each with its own administrative boundaries. The challenges of regeneration in such an organisational and administrative maze have been characterised as 'complex' (Hall, 2009), 'diverse' (Brownill and Carpenter; 2009), 'disconnected' (Poynter, 2009) 'hyperactive' (Catney et al., 2006) and 'demanding' (Allmendinger and Haughton, 2009b).

The history of regeneration in the area goes back many decades, though it was the establishment of the flagship London Docklands Development Corporation by the first Thatcher government in 1980 that kick-started its redevelopment after years of decline. The publication of *Sustainable Communities* (ODPM, 2003a), *Creating Sustainable Communities: Delivering in the Thames Gateway* (ODPM, 2005b) and the *Strategic Framework for the Thames Gateway* (DCLG, 2006g) revealed the government's intentions, including funding arrangements. The *Sustainable Communities* plan established ambitious targets for the Gateway area, notably the delivery of 160,000 new homes and 180,000 jobs (later raised to 225,000) by 2016. The overall aim was to create sustainable communities through spatial planning (ODPM, 2003a). The challenges of, and justification for, a spatial approach in the Thames Gateway were clear: with a congested institutional landscape and significant existing public expenditure in the area, coordination of future development would be critical. The House of Commons ODPM Housing, Planning, Local Government and the Regions Committee report of 2003 (HoC ODPM, 2003a) had pushed heavily for a more integrated and coordinated approach to planning in the area and a greater steer for central control over the multitude of departments, bodies and authorities.

There have been a plethora of studies and evaluations of regeneration in the Thames Gateway (e.g. NAO, 2008; HoC ODPM, 2003b; Imrie *et al.*, 2009; Brownill and Carpenter, 2007), though few explore the spatial planning component. Nevertheless, analysis of the experiences so far highlight some important issues around the notion and practices of spatial planning (Allmendinger and Haughton, 2009a; Haughton and Allmendinger, 2008; Haughton *et al.*, 2010).

First, spatial planning can be in tension with the requirements of regeneration and development. The 'lock-in' of different bodies and strategies required as part of the spatial planning process can work against the need for flexibility and a 'light touch' to adapt to changing circumstances and unforeseen events. A senior politician in the Greater London Authority commented:

> ... you need a framework which is sufficiently loose that it allows unexpected things to happen ... Stratford City ... we got very excited a while ago ... we said here is the master plan ... this is what's going to happen ... [however] the market will decide how most of that will happen ... City East is arguably a strategy that was dreamt up by Lord Rogers about two years ago ... But it doesn't feature in the London Plan. What happened was ... it's a good example of where people say if we look again at this area perhaps we will find another node for growth which wasn't described in the London Plan, but maybe is valid ... What does that tell you about the balance between prescription and facilitation? I don't know ....
>
> *(Haughton et al., 2010: 211)*

This balance between certainty and flexibility is inherent within any planning system, and with the UK approach the emphasis has varied through time. Spatial planning privileges coordination and certainty, while previous regimes in the area – specifically, the London Docklands Development Corporation – were concerned more with speed and certainty. However, an added dimension in the Thames Gateway was the initiation of a new suite of plans and documents from 2004, including the London Plan. The needs of spatial planning were taken seriously by those involved, who were also tasked with preparing a new suite of development plan documents and doing so in a vaguely defined, 'spatial' way. Spatial planning also meant working beyond formal boundaries, which created a complex landscape of documents and strategies that worked against delivery:

> The department [DCLG] has encouraged the development of several forms of partnership at regional, sub-regional and local level to help co-ordinate investment across the Gateway .... the complexity of the decision-making and delivery chains makes it difficult for potential investors, developers and Government itself to understand the programme and integrate investment as a whole.
>
> *(National Audit Office, 2007, p. 5)*

In the particular complex circumstances of Thames Gateway, the approach to spatial planning was felt by some to be part of an impediment to regeneration rather than a solution to it. As a housing developer noted:

> I'm firmly of the view that the government needs to take an axe to the whole of this and create a single body for the Gateway ... which is a UDC in the true sense of the word ... whose sole objective is delivery of the government's objectives for the Gateway.
>
> *(Haughton et al., 2010: 205)*

This view was not confined to the private sector. One partnership agency commented: 'There is no-one in overall control ... Different actors are doing different destructive things ... It's going to be guerrilla war again ... Just because it's screwed up ... there's so many loose ends ... nobody's coordinating it' (ibid.). A common view is that the Thames Gateway is 'over-planned' and an example of 'congested governance' (Sullivan and Skelcher, 2002).

The second lesson of spatial planning in the Thames Gateway shows up the consequences of the vagueness of the concept. Spatial planning does not privilege particular scales but, in an echo of New Labour rhetoric, focuses upon 'what works'. While flexible, this has led to confusion and overlap. The question of where policy derives from and the source of legitimacy are unclear. The elected status of the London mayor has provided a platform on which to intervene legitimately in borough planning matters. Since 2008 the mayor's powers to call in applications of strategic importance have increased (Tewdwr-Jones, 2009). The London Plan has been welcomed by many as providing a strong, strategic framework for planning in the capital, while others, particularly the boroughs, have complained that it is too detailed and leaves little room for local discretion and flexibility (Haughton *et al.*, 2010). The upshot is a clear tension between, on the one hand, flexibility and the need for networked forms of governance and, on the other, the need to have a strategic overview (Brownill and Carpenter, 2008).

Poynter argues that the successful bid to stage the 2012 Olympics and the focus upon the Olympic Park and Lower Lee Valley has provided a contrast in governance styles with the wider Gateway. National concerns with delivery are dominating more local, partnership-led approaches: 'The achievement of a "successful" [Olympic] Games appears to rest upon the capacity to sustain the public rhetoric of partnership while the national increasingly dominates local interests' (Poynter, 2009: 139). One senior London politician in the Gateway commented that:

> There are clearly tensions ... one of the biggest tensions is the territoriality tension ... the boroughs will see themselves being sovereign over the individual land use issues and they will expect the spatial strategy to set themes and general direction for their work ... As a consequence of that, any sane mayor will want to exercise their authority in a more detailed spatial fashion

> ... so there is a tension between the aspirations of the London Plan which is sort of general and enabling and the desire to be prescriptive ... and the desire of the boroughs to maintain their own turf.
>
> *(Quoted in Haughton et al., 2010: 199)*

Despite criticisms from the Public Accounts Committee (HoC PAC, 2007) and the National Audit Office (NAO, 2007), there has been little movement from government to resolve this tension in the Gateway area (Allmendinger and Haughton, 2009b).

In an echo of the findings of others (e.g. RTPI, 2007), few agencies or individuals involved in regeneration in the Gateway mentioned the Community Strategy or Local Strategic Partnership as an important element in their work. This is, perhaps, part of the reason for the, at best, patchy sectoral coordination. Coordination between development strategies and health, for example, has been good, while that involving education has been poor (Haughton *et al.*, 2010). But this is not necessarily the fault of spatial planning, as coordination is a two-way process. It is clear that some sectors have embraced coordination more than others, particularly within central government, highlighting a fundamental and exogenous factor in the success and impact of spatial planning.

One significant outcome of the above has been the emergence of 'shadow' or informal spaces and processes, along with 'fuzzy boundaries'. The formal spaces, processes and requirements of plans and strategies derived from legislative requirements are being supplemented by more informal plans and strategies based around functional planning areas. Such areas are often related to large sites or transport or infrastructure hubs and were used either to avoid the formal processes, timescales and consultation requirements of, for example, LDFs or to focus attention and action upon specific and deliverable outcomes. This is a creative use of discretion and, arguably, within the spirit of spatial planning. One view is that the emergence of such shadow processes and plans could be interpreted as planners and others passing judgement upon the new, post-2004 system of planning and the functional requirements of regeneration. Separating spatial planning as a creative, place-making process from the requirements of formal plan- and strategy-making is a common element in both the Cambridge and the Thames Gateway approaches.

Conclusions

If we separate spatial planning from plan-making and explore the practices of planning as they arose, it is clear that spatial planning is not emerging as a new paradigm (see also Newman, 2008, and Haughton *et al.*, 2010). The moves by the coalition government in 2010 seem to point to a significant reorientation away from the spatial and back to towards a more regulatory approach (Chapter 8). There would appear to be a number of reasons for this. First, spatial planning is hardly new. The introduction of Structure Plans in 1968, the expected reorganisation of local government along sub-regional lines, the need to incorporate regional strategic plans and city-region structure plans, and the requirement that these would be embedded in local authority-wide corporate management plans provided a legal context for and planning culture of multi-spatial, coordinated plans.

Second, local authorities 'picked and mixed' the elements of spatial planning that best suited them, their circumstances and their development vision. Struggling to understand the new system of development planning, the vague notions of spatial planning and the multiplying requirements and objectives for planning, many simply retreated into 'what works'. Whether this could be labelled spatial planning is moot. There is also another point, which concerns the damaging conflation of spatial planning with the 2004 Act. Understanding that the two are distinct helps explain many of the problems and issues that have subsequently arisen in planning. It was not the intention of the 2004 Act to make the system more complex, yet complexity has grown as more and more objectives and procedural requirements were layered onto RSSs and LDFs, including revisions to the process in the Planning White Paper (DCLG, 2007a):

> The LDF system is high maintenance, and the demands on content are growing; having got flood risk assessments, sustainability appraisals, equality impact assessments, health impact assessments and strategic housing market area assessments, to name but a few, under their belts, policy planners are now getting stuck into infrastructure plans and will be rubbing their hands at the prospect of the local economic assessments proposed in the Sub-National Review.
>
> *(Watson and Crook, 2009: 124)*

Despite the 'buy-in' of the DCLG to spatial planning, the Treasury-led subnational review of planning is underpinned by a more regulatory land-use allocation perspective of planning, highlighting a severe mismatch between the PPS12 view of spatial planning and the Treasury's perspective of more regulatory planning being a delivery mechanism for sustainable economic development (Baden, 2008).

Given the above, was it any wonder that planners were bewildered and sought to escape into soft spaces and processes? Such 'soft' solutions were encouraged by government through the emphasis on sub-regional planning (e.g. the requirement for functional housing markets and travel to work assessments as part of the evidence base of LDFs) and the introduction of Multi-Area Agreements and other incentives to plan outside and alongside 'hard spaces', as in the case of Cambridge. In effect, this amounted to the centre condoning if not encouraging the development of alternatives to its 'formal' system.

Ironically, spatial planning as envisaged by some could actually have done the profession and practice of planning more harm than good. The conflation of spatial planning with the post-2004 system of development planning meant that the dissatisfaction with the latter coloured views of the former. In a thorough and comprehensive survey of attitudes towards planning modernisation, Clifford (2007, 2009) highlights both the broad support for planning reform, spatial planning and the new system of development planning and the widespread dissatisfaction and disenchantment with the process and outcomes. Clifford found that the feeling among planners in the public sector was that the new system was labour intensive, overly focused upon process, under-resourced and too complex. As one planner put it:

> What we're suffering from really is this Government not being joined up in its thinking ... we find ourselves, at the local level, trying to resolve issues that really should have been sorted out elsewhere. For example, disputes between government agencies and the Government.
> (Quoted in Clifford, 2009: 174)

Dissatisfaction with the spatial approach can also be found elsewhere:

> There is a major failure to take account of neighbouring communities (administrative boundaries still rule); there is a lack of integration in terms of the link between plans and investment; and there is a long way to go to engage effectively with communities.
> (Goodstadt: 2009: 8)

For those who focused upon spatial planning as the *primus inter pares* of strategies within local authorities, there was also disappointment. There has never been the relationship between LDFs, Community Strategies and Sustainable Community Strategies envisaged by some. Where such coordination was achieved, it was more likely to be secured and delivered through old-fashioned financial incentives, as in the case of Cambridge.

Notwithstanding the above, the final point, and the main reason why spatial planning is likely to be a short-lived phenomenon, concerns delivery. For some, such as the Conservatives, the planning system in the first decade of the twenty-first century was delivering too much growth and housing in local communities that simply didn't want either. Their solution, discussed in Chapter 8, has been radically to shift the balance within planning in favour of communities. However, for those wishing the more delivery spatial planning was also flawed. Those advocating spatial planning as a solution to the concerns of New Labour focused too much attention upon delivering spatial plans and planning rather than delivering development. The traditional Cinderella of planning – development control (or development management, as it was later labelled) – was largely ignored. There was an assumption that spatial plans – the RSS and the LDF – would lead to perfect implementation and that development control decisions would 'flow' from the plan. Indeed, Section 38 of the 2004 Act was introduced to re-emphasise this link and further to underpin it legally. The problem, as the government slowly began to realise, was that communities were less than enthusiastic about development and would use the planning system as a whole to thwart it. The annual Saint Index of public attitudes towards development (Saint Consulting, 2007, 2009) highlighted the scale of the issue. This animosity, particularly in the south of England, where much of the growth was concentrated, was combined with the deployment of vague and positive narratives such as 'sustainable development' and 'urban renaissance' in national planning guidance to create an imbalanced system of national and regional growth promotion and local resistance. Despite the link between the two pillars of planning – development plans and development control – conflict and resistance emerged once a proposal was submitted, as described in the King's Cross case in Chapter 4. Reforms to development control focused largely on the process through the introduction of the Planning Delivery Grant (see Chapter 6), as if the conclusions and decisions had been determined elsewhere through the plan. The result was that development control became highly politicised and the weak link in the delivery of spatial planning and development.

6

HITTING THE TARGET AND MISSING THE POINT

Speed in planning decisions

Introduction

Development control has long been the Cinderella of planning, largely ignored or overlooked and wrongly perceived as being narrow, regulatory and technical. Such a view tends to approach it from a management perspective and see it as being about process. Criticism of the speed of decision-making has been a consistent theme and has emphasised the idea of development control being a process that largely endorses or refuses development proposals by comparing them to the outcomes of decisions taken elsewhere. The Dobry report in the mid-1970s, for example, explored the reasons for delays in development control and recommended better management of the process and applications (Dobry, 1975). The Conservative government from 1979 also introduced a range of measures aimed at making decisions quicker. The concerns of successive governments and the largely unsuccessful attempts to speed up planning in general and development control in particular represent a fundamental characteristic of UK planning. The separation of plan-making from decision-taking in the postwar system was envisaged as a way of separating strategic from detailed issues. This separation has always had advantages and disadvantages. The much vaunted flexibility and discretion, on the one hand, has to be weighed against speed, cost and complexity, on the other (see Haughton et al., 2010: ch. 2). As a result there have also been successive attempts to 'bridge the gap' between the two without actually moving wholesale towards the zoning-based approach of combined plan and permission more common in continental Europe and North America. The 'plan-led' approach introduced in 1991 and reinforced in 2004 sought to link plan and permission with a presumption in favour of the development plan. The primacy of the plan, among other material considerations, was not only an attempt at minimising local discretion but also, through the requirements for approval of development plans, a way of centralising control (Haughton and Allmendinger, 2009b; see also Chapter 1).

The attempt to provide a continuum between plan and permission through the plan-led approach has been largely unsuccessful in speeding up decisions or providing greater certainty. However, one unintended consequence of this 'sticking plaster' approach was to downgrade the role of development control in the minds of local authorities and planners. The visionary and strategic role of development planning was played up and contrasted with the largely bureaucratic and technical exercise of development control. Attempting to remove discretion, downplay the separation of plan from permission, and highlight the strategic and spatial nature of planning naturally undermined the development control function, and planners acted accordingly, preferring to work in high profile and 'creative' fields such as regeneration, spatial planning and town-centre management, if not to move to the burgeoning private sector.

However, problems of quantity and quality in staff were by no means the only problem. Development pressures and activity have increased against the backdrop of a stalling post-2004 development plan system. By 2009 only 20 per cent of local planning authorities had a core strategy in place (Watson, 2009). One outcome was an application or appeal-led approach rather than the desired plan-led system. Another was the increasing politicisation of development control, as it became the main arena for mediating conflicts over future development that were originally envisaged to be part of the development plan (see Chapter 4). At the same time, requirements for supporting information and government objectives ballooned (Killian and Pretty, 2008) while local opposition to development increased (Saint Consulting, 2007). The upshot in an area of planning that has been traditionally under-resourced was as predictable as it was inevitable: 'we found wide frustration on the part of applicants, councils, interest groups and consultees about how the current applications process was operating and, to varying degrees, about how slow, unpredictable, and costly it had become for all involved' (Killian and Pretty, 2008: 3).

New Labour was not immune to the widespread criticisms and suggestions of how to reform and speed up planning. As far back as 1996, Keith Vaz, the then shadow minister for planning and regeneration, stated: 'there is still a culture of delay within the planning system' (Vaz, 1996: 12). This attitude continued into government. Echoing the view of the CBI, among others, Lord Falconer, minister of housing and planning, in a speech to the British Urban Regeneration Association on 14 February 2002, said that 'the UK loses out because the glacial pace of change reduces our global competitiveness'. Both the obsession with speed and the assumption that slow development control processes inhibited economic growth have been common across governments and through time. Nevertheless, the responses of different governments have varied. More active measures have included attempts at deregulation through, for example, extensions to the General Permitted Development and Use Classes orders. Experiments such as the limited introduction of zoning approaches in Enterprise Zones and Simplified Planning Zones were periodically introduced. Other, more 'passive' measures, such as time-related performance targets, were aimed largely at 'naming and shaming' local planning authorities and identifying those where intervention by government might be justified. Despite these changes, the 'problem' of delay in development control did not go away and, well into Labour's second term, actually got worse as application numbers rose and performance deteriorated.

New Labour's approach broke with the 'passive–active' model and began to address what local planning authorities, with the backing of agents and developers, had long argued were the more fundamental issues of profile and resourcing. The most obvious way of improving resourcing was to increase fees. However, agents and others successfully argued that performance in most local authorities could not justify substantial fee increases. To improve performance a 'carrot and stick' approach was developed. Local authorities that met targets would be financially rewarded while those that did not might face intervention, including, ultimately, losing control over the service. Once overall performance had been improved, then application fees would be raised to ensure that the full cost of dealing with an application fell on the applicant. The second prong of this approach sought to address the more fundamental issue of the separation of plan from permission through attempting to strengthen the link in decision-making between the two (Allmendinger, 2006; see also Chapter 2). Planning was not unique in the shift towards the increasing use of performance metrics to assess and improve the performance of local authorities (McLean et al., 2007).

The combination of these elements amounted to a 'commodification' of planning, based around:

- a locally managed system of review and restructuring;
- centrally driven national performance indicators;
- rewards and sanctions;
- training, organisational development (e.g. Improvement and Development Agency (IdeA), the Planning Advisory Service, Planning and Regulatory Services Online (PARSOL), initiatives such as Advisory Teams for Large Applications (ATLAS)) and support for management change (e.g. e-planning);
- the reduction of local discretion in development control decision-making through the strengthening of the link between plan and permission and selective re-regulation, including Local Development Orders, extensions of permitted development and increases in the scheme of planning officer delegation.

This chapter focuses upon the use of rewards and sanctions as part of national performance indicators and explores the implications and impacts. The broad argument is that planning in general and development control in particular were subject to the wider obsession of New Labour with performance indicators and targets as part of the broader approach to 'management by numbers' (Hood, 2007). As I conclude, this approach failed to recognise or chose to ignore the fundamental discretionary characteristics of development control and its local institutional context. Development control managed, on the whole, to hit the targets, allowing the government to point to performance improvements and 'modernisation'. However, the actual outcomes were far less clear cut.

Performance management in local government and planning

Morphet (2008) rightly concludes that the performance framework for local authorities expanded considerably under Labour. Despite the expectation of many in local government that performance measurement and indicators initiated under the previous Major administration would be relaxed, Labour expanded the scope and detail of such regimes soon after coming into office. While the introduction of Best Value diluted elements of competitive tendering, it confirmed that Labour and Whitehall still distrusted local government and were prepared to use the Audit Commission in a scrutiny role, particularly when Comprehensive Performance Assessment was introduced in 2003 (Hood, 2007).

As Carmona puts it, the development of performance indicators in English planning graphically illustrates this concern for speed above all else (2003a, 2003b). The first concerted effort to measure performance came, unsurprisingly, during the first Thatcher administration. In 1979, government decided to collect data on development control performance relating to the proportion of decisions within eight and thirteen weeks. In 1982 it then indicated that local authorities should be aiming to determine 80 per cent of applications within eight weeks (Carmona, 2003a). The publication of planning and other indicators by the newly established Audit Commission during the 1980s, however, lacked teeth: local authorities simply argued that poor performance figures were a function of resource constraints or the local decision to focus upon development quality rather than speed. Some developers and applicants began to act strategically in those authorities that performed poorly by submitting two identical applications and appealing against non-determination of one after the passing of eight weeks, a practice now halted.

Throughout the 1980s the priority of government was speed of decision-making. In 1992, however, beleaguered local authorities could point to the influential Audit Commission report *Building in Quality*, which highlighted a range of consequences of this obsession with speed. The report argued that the focus upon the figure of eight weeks discouraged negotiation and ignored quality of outcome. For the first time local authorities had support for their argument that targets needed to be more sensitive to different types of development: 'The quality of outcomes is more important than the quality of process, because buildings will be seen long after memories of the decision process have lapsed' (Audit Commission, 1992: 19). The government response was not supportive of the Audit Commission's position (Carmona, 2003a), though the tide against such rather uni-dimensional indicators was turning.

The anti-local government attitude of the Thatcher administrations gave way to a more subtle strategy under Major during the 1990s. After a decade in which powers were taken away from local government and vested in a range of institutions, local government was again recognised as a key player in service delivery. While, in principle, this was a more 'bottom-up' approach, the centre still controlled the process, most notably by the way in which resources were allocated through competitive funding regimes. One of Major's (in)famous policies came in the form of the Citizen's Charter, launched in 1991 and initiated from 1992 with the aim of improving public services through the publication of service standards, the right of redress, performance monitoring, and penalties for public services that did not meet standards. Citizen Charter Performance Indicators were published by the Audit Commission for every local authority, which in turn was required to publish performance against targets. The inevitable outcome was league tables of local authority performance that focused largely on speed and efficiency. While the tables were heavily criticised at the time for their choice of indicators and the costs of collecting and publishing data, two more fundamental issues concerned the reduction of complex processes into simplistic indicators and the imposition of nationally derived performance expectations into local political processes. The notion that each local authority should be accountable to the electorate for its performance was, in part, broken. The electorate now had access to information that allowed them to compare performance with that of other authorities, even though that performance may have reflected local rather than national priorities.

The Audit Commission published additional guidance which conceded that, in a wide range of circumstances, the Citizen's Charter planning indicators might be misleading (Audit Commission, 1994). Applications relating to listed buildings and those in conservation areas would be more complex and time-consuming, and where there was no up-to-date development plan there would inevitably be a large number of departures. This provided authorities with ready-made explanations for performance that deviated significantly, effectively undermining the point of the exercise. However, more importantly, there were early indications of strategic behaviour on the part of authorities to meet targets, including a reluctance to negotiate and improve proposals, as refusing an application amounted to a 'decision' within the time limits (McCarthy and Harrison, 1995).

Following Labour's election victory in 1997, the concept of Best Value was introduced in the Local Government Act 1999 and replaced the Citizen's Charter. Best Value aimed to introduce the notion of continuous improvement in local government and was based upon ninety Best Value Performance Indicators (BVPIs) to be collected and audited by the Audit Commission. BVPIs for the planning function of local planning authorities were:

- BV106 – Percentage of new homes built on previously developed land;
- BV107 – Planning cost per head of population;
- BV108 – The number of advertised departures from the statutory plan approved by the authority as a percentage of total permissions granted;
- BV109 – Percentage of applications determined within eight weeks;
- BV110 – Average time taken to determine all applications;

- BV111 – Percentage of applicants satisfied with the service received;
- BV112 – Score against a checklist of planning best practice.

Local authorities were allowed to set their own standards for each indicator, though they had to have regard to the national targets for the reuse of brownfield land in the processing of planning applications (Carmona, 2003a). The emphasis in BVPI was, again, largely on process rather than outcomes and, despite the intention to allow local authorities to set their own standards, the government reserved the right to intervene if authorities did not meet what were felt to be minimum requirements. Pressure on the government to raise performance also led to the indicators being changed from 2002/3 to include, for the first time, a distinction between major and minor as well as commercial and residential proposals. This allowed differential targets for speed of decision-making to be introduced, overcoming the crude averaging of all figures to do with application processing and removing the incentive to focus upon the higher number of smaller applications to boost overall performance.

Resourcing planning

This rather simplistic central concern with speed in development control continued as a theme in the 2001 Planning Green Paper (DTLR, 2001a). According to Stephen Byers, secretary of state for transport, local government and the regions,

> The Green Paper proposes a much stronger emphasis on customer service, including delivery to business. New targets for processing planning applications by local authorities will distinguish business from householder applications in order that all applicants should have a clear and realistic expectation of the speed of decision. We propose that delivery contracts should be agreed between local authorities and business for the biggest planning applications that would include an agreed timetable for reaching planning decisions.
>
> *(Hansard, 12 December 2001, vol. 376, cc881-2W)*

The Green Paper set out the government's thinking on development control performance as target driven – a view which suggested that there was no fundamental impediment to improved performance. This underlying thinking had evolved little since the successive failed attempts to achieve faster decisions throughout the 1980s and 1990s. In its assessment of the Green Paper's proposals on performance, the House of Commons Transport, Local Government and the Regions Committee, following evidence from a variety of sources, questioned the assumption that improved performance was simply a matter of will power:

> The Committee was astonished by the lack of attention to the most obvious problem facing the delivery of an effective planning service, namely its

under-resourcing. There is a shortage of professional and experienced planning staff in most local authorities, low morale and a recruitment problem. Ministers' obsession with shaming authorities with poor performances, measured largely in terms of speed rather than quality of decisions, has no doubt contributed to this. Meanwhile, local authorities divert money away from planning to other more politically attractive uses.

(HoC TLGR, 2002: paragraphs 211–12)

As the committee went on to point out, the Green Paper went much further than simply introducing performance targets. A new system of development plans would be introduced with new processes, requirements and relations between plans:

> The Government accepts that its proposed new system will take more planners to operate than the current one, but has no serious proposals in hand to train and attract staff even to fulfil current requirements. We have no hesitation in recommending that significantly more money and staff should be ploughed into planning long before any major revision of its practices is contemplated. We suspect that some at least of the problems identified would diminish or dissolve if more and better qualified staff were in post to address them.
>
> *(Ibid.)*

With support from planners and developers, who tended to agree with the committee's analysis, the then Department of Transport, Local Government and the Regions commissioned research to explore how the planning system was resourced. The study, whose 'overwhelming finding is that resources have declined significantly over the past five years and performance has generally worsened, albeit in different functions in different Authorities' (DTLR, 2002a: paragraph 8) stated that:

- the level of cost recovery on planning application fees had declined. An increase of at least 14 per cent for districts and unitaries was justified to reflect inflation and achieve cost recovery;
- planning authorities typically handled householder applications at a cost of around £200 per application, compared with a fee of £95;
- the overall picture was of a system that operated efficiently but at a net loss.

(Ibid.: paragraph 21)

This link between performance and resources was backed up in evidence to the House of Commons committee, which concluded that agents and applicants were willing to pay more for a better service (HoC TLGR, 2002: paragraph 200). The outcome was a range of initiatives related to the resourcing of planning. Section 53 of the 2004 Act, for example, made major changes to the ability of local authorities to charge for planning work. The possibilities of charging for appeals and above inflation fee increases were also floated by the government. However, the most significant change in relation to performance was the announcement of the Planning Delivery Grant.

The Planning Delivery Grant (PDG) was introduced for the financial year 2003/4 with the aim of improving the performance and resourcing of planning authorities. It aimed to reinforce the importance of existing development control performance targets, specifically BVPI109. Between 2003/4 and 2007/8, around £68 million per year was allocated to local planning authorities on the basis of performance. At the same time the Planning Advisory Service (PAS) was introduced to help authorities improve performance and the Advisory Team for Large Applications to assist Authorities (ATLAS) was established to help support authorities dealing with large applications.

Within a couple of years performance seemed to have improved significantly. Published figures showed a dramatic improvement, with 67 per cent of authorities dealing with major residential applications within thirteen weeks in 2007–8, compared with 37 per cent in 2002–3 (NAO, 2008). This improvement needs to be seen against the backdrop of increasing application numbers during the period. The government's review backed up the idea that performance had improved as a result of PDG: 'PDG has certainly encouraged authorities to look at their development control performance whether this has been poor or not, improving to avoid being or becoming a Standards Authority and to maximise future grant' (ODPM, 2004e: 8). Early evidence was that local authorities were improving performance by changing internal management processes, raising expectations of what was expected when applying for planning permission, and refusing to register incomplete applications. Had PDG succeeded where decades of targets around speed of decision-making had failed?

Playing the game? Housing supply in the South-East

Despite published performance improvements, there has always been a significant degree of scepticism that development control performance improved under Labour (CBI, 2005). There was a strong feeling that securing planning permission, as opposed to obtaining a decision, was no easier or quicker under the BVPI/PDG regime. One issue was that local authorities were felt to be acting strategically in manipulating the process to give the appearance of improved performance. Another was that the actual measures used by government were, at best, partial: 'Many users argue that measuring the time from registration/validation to decision does not reflect the totality of their experience, which would include, in particular, pre-application discussions and discharging pre-start conditions' (Killian and Pretty, 2007: 2). Performance measurement under Labour, as under previous administrations, focused upon the period between registering an application and issuing a decision. However, this is a small proportion of the total time that an authority may take in dealing with a proposal to develop a site. This measure ignores or does not account for repeat applications on a site, either when an application is refused or when a developer needs change and a modified scheme is submitted. This amounts to a difference between an application-based approach and one based upon sites. A site-based measure is concerned with the total time taken to secure an implementable planning permission.

The general concern of developers and others was that an application-based approach was partial and simplistic. Figure 6.1 highlights the main difference between application- and site-based performance measures of the development control process. The top line is a simplified development control process. Performance measurement was focused upon what is termed the formal application process. However, a site could have multiple applications, all of which could be determined (i.e. lead to a decision) within the target time but none of which could amount to an implementable scheme. Multiple or sequential applications or modifications to extant permissions are not uncommon, particularly for larger or more complex sites. As market conditions and owners/developers change, so the scheme may evolve. Equally, much of the process of securing an implementable permission lies outside of the formal process. To capture fully the time taken to secure an implementable planning permission there is a need, therefore, for a site-focused assessment of delay that can account for multiple applications for similar schemes (both approval and refusal) and the time taken to negotiate in the informal process.

This site-based approach to the measurement of performance in development control and the impacts of PDG and Labour's modernisation agenda formed the focus of research undertaken in 2007–8 (Ball et al., 2009). The study undertook to compare application- and site-based measures of performance and the extent and implications of strategic behaviour on the part of local authorities linked to the use of PDG.

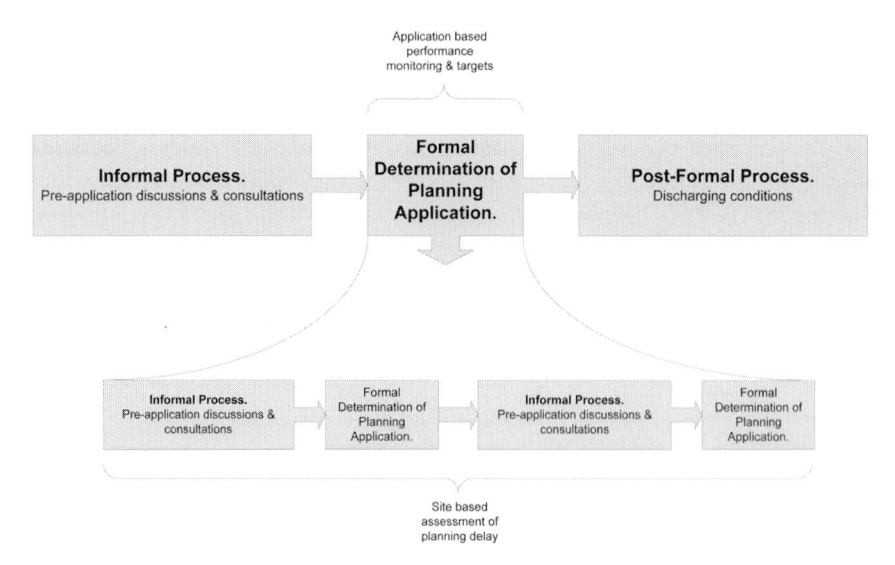

FIGURE 6.1 Application- and site-based performance measures of the planning process.

Data were analysed from a sample of residential developments successfully completing the planning stage of development in 2006 for eleven local authorities in Berkshire, Hampshire and Oxfordshire, and two measures of planning process time were estimated for each site. The first, termed net planning days, identified the total amount of time a planning permission was pending – i.e. the sum of the dates between an application or appeal being lodged and a decision being sent out (overlapping dates were not double-counted). Appeals were included within this time frame. The second, termed gross planning days, identified the full period from the first planning application to the final approval of the last planning application made with respect to the development. This second measure includes the time when the developer has no outstanding applications for the site but was either preparing a resubmission or strategically holding on to the land for market, land banking or expected changes in planning environment reasons. This period corresponds to the site-based approach in Figure 6.1.

Summary data for these times, measured in weeks, are shown in Table 6.1. The median value for time in the planning system for a residential development was almost eleven months, with the median total time lasting another eighteen weeks. The measured standard deviations are also high, so many sites were delayed for far longer: 41 per cent took over a year to be processed, 17 per cent took over 100 weeks, and 6 per cent took longer than 150 weeks.

At an individual authority level there is clearly a large discrepancy in time using the application- and site-based measures (Table 6.2). These site-based times are clearly far longer than the thirteen-week target set for individual planning permissions for major applications. Only 8 per cent of sites were processed within that target and only a 20 per cent within twenty weeks. The number of weeks to process a site through from the date of the first application to final approval typically runs from nineteen weeks for the fastest local authority to seventy-eight weeks for the slowest, or, on a per dwelling basis, from 1.7 weeks for the fastest to 3.1 for the longest. The difference between the two is over four times on a site basis, although somewhat shorter on a per dwelling basis.

The potential influences on the time it takes to progress a site through development control may be highly varied, especially in the discretionary-based UK approach. However, in practice, the vast majority of schemes are evaluated on a relatively limited set of criteria. The main influences used in the sample included

TABLE 6.1 Development control time for individual sites

Time in weeks	Total planning days	Net planning days
Mean	84	58
Median	62	44
Standard deviation	72	44

Notes: 180 sites from local authority survey.
Net planning days = total time with outstanding planning applications or appeals.
Total planning days = time from first application to final approval.

TABLE 6.2 Planning process time by local authority and development control performance, 2006

	Site-based measure		Application-based measure
	Net number of days in planning	Net number of weeks in planning	Percentage of major applications determined within 13 weeks, Oct–Dec 2006
Reading	133.0	19.0	62
Eastleigh	143.5	20.5	77
Woking	202.5	28.9	65
Hart	246.0	35.1	80
White Horse	258.2	36.9	92
Wokingham	276.8	39.5	65
East Hants	304.6	43.5	70
Winchester	331.3	47.3	46
Slough	349.4	49.9	93
Guildford	364.9	52.1	52
Basingstoke	434.7	62.1	69
West Berks	454.5	64.9	85
Portsmouth	546.2	78.0	70

Source: for final column www.communities.gov.uk/documents/planning and building/pdf/321235.pdf.

site characteristics (e.g. site area, number of units, density, greenfield or brownfield designation), characteristics of proposed buildings (e.g. building mix), local authority characteristics (e.g. committee structure and scheme of delegation) and developer characteristics (e.g. firm size, use of local agents).

The research also explored the causes of planning delay through the identification and quantification of specific characteristics determining the amount of time required to evaluate a development project and award it planning permission. The overall results indicate that a relatively limited number of variables can explain a substantial part of the total. This implies that planning process times are systematically influenced by a handful of core factors rather than being random outcomes within a standardised planning framework. Put another way, the individual case approach fundamental to the UK method of development control seems to have a substantial and variable impact on the times taken to process housing development sites. There are six main points from the analysis.

- The most consistently important variables were the local authority dummies, all of which were strongly significant and explained much of the variance. Project characteristics, perhaps surprisingly, seem to have had little influence on planning process times, with only project size being of significance. The type of developer was also influential to a degree.
- Project size is important in capturing the influential factors determining the length of time taken by development control processes. Larger projects took

longer to process and larger developers experience longer planning times for their projects.
- Housing association developments experienced less time in the planning process than others. The widely held notion that larger developers have an edge through their greater resource and skill base seems to be rejected by these results, but the notion that planners favour social housing and local builders is not ruled out.
- Scale economies in processing planning applications were indicated. Both medium-sized and large projects saw significant reductions in time taken per dwelling, with the largest schemes gaining the biggest time saving per dwelling. So, even though larger schemes take longer to process through the system on a site basis, there is an overall reduction in the time required per dwelling.
- In a handful of authorities both developer and site characteristics were also significant in explaining delay. In these authorities large sites were associated with more planning applications, probably because of their relative complexity and frequently controversial nature and because developers may wish to change the dwelling mix as the development comes closer to fruition.
- Both the brownfield and the smaller site variables were also significant. This suggests that inner-city developments were often strongly contested by developers and planners. Detailed examination of the documentation indicated particular disagreement between developers and planners over design, density and parking matters. Planners also seemed concerned about the impact of even small new developments on congestion. Developers had a propensity to alter configurations, especially to change the number or types of apartments or to improve a development's marketability – for example, by trying to enhance parking facilities.

The overall argument from the research is that the time taken to secure planning permission arises from a variety of local 'institutional' factors, such as the culture of the planning authority; local politics and their influence on planning committees/decision-making processes; how well authorities signal preferences to developers; how easily the authority may change its mind when developers put in further applications or appeals; and the propensity of developers in a locality to challenge or negotiate with the planning authority through the appeal process.

The final element of the research looked at attitudes towards development control performance from local authority planners and housebuilders through a survey and semi-structured interviews. The survey found that 82.4 per cent of respondents considered performance targets had led to increases in rejections, and this perception is confirmed by government statistics: approval rates for all applications fell from 88 per cent in 1998/9 to just over 82 per cent in 2007/8. The same proportion of housebuilders felt that requests to withdraw applications had also increased, and, again, there is confirmation of this from the government, with withdrawals increasing from just over 4 per cent in 1997/8 to around 6.5 per cent in 2006/7. There was a strong feeling among housebuilders that these and other consequences arose directly from the introduction of performance targets (Ball *et al.*, 2009).

It is only fair to stress that such strategic behaviour was not confined to local authorities. The research also highlighted strategic practices on the part of applicants. One issue that concerned many developers is the desire to overcome local resistance to any form of development by using the appeals process. Some developers would submit a proposal that constituted the absolute minimum of what would be acceptable in order to get the application registered. When a local planning authority requested additional information the applicant would stall, forcing the authority to refuse the application. The inspector at the subsequent appeal would then be furnished with all the required information and be able to make a decision that the local planning authority were never in a position to make.

Rewarding delivery: the Housing and Planning Delivery Grant

Concern over the mismatch between published performance in development control and experience went wider than simply value for money and efficacy. There was also a growing concern that government was focusing upon the wrong elements of the system if it was to achieve its wider objectives around housing supply. The first Barker report (2004) had argued for the need to provide better incentives for housing delivery through PDG, a theme which was later echoed in a report by Sir Michael Lyons (2006). The final evaluation of the Planning Delivery Grant had picked up on both this and the wider dissatisfaction with PDG and explored the possibility of changing the latter's emphasis from rewarding processes to rewarding outcomes. The government accepted that local authorities had begun to see process-related performance as a means rather than an end, leading to perverse outcomes. A consultation paper on a possible shift in the focus on PDG was issued in 2006 (DCLG, 2006i) and sought views on how incentive funding to local authorities could be taken forward. In recognition that PDG had led to strategic behaviour on the part of some local planning authorities, the consultation on the Housing and Planning Delivery Grant (HPDG) set out the government's desire not to see this repeated:

> first and foremost we would expect local authorities to carry out their place-shaping role duties responsibly in the interests of the whole community. We would not expect them to seek outcomes that are to the long-term detriment of the community simply for the sake of short-term grant.
> *(DCLG, 2007c: paragraph 13)*

Responses on the shift in PDG from process to output were mixed (DCLG, 2007c). A repeated concern was the perceived danger of incentivising quantity over quality. However, two significant issues were not addressed in the paper.

First, it was not clear how the new scheme could reward authorities for delivering something that was beyond their control. Authorities do not have powers to compel building to occur once permission has been granted, only those to enforce completion of unfinished developments. Development could be thwarted for a variety of reasons after permission has been granted and local authorities would have no control over this. The subsequent collapse in housebuilding from 2008 highlighted the degree to which housing development did not necessarily follow once planning permission had been secured, however quickly. Second, housing markets and delivery operated in areas that did not normally correspond to local authority boundaries. In other words, housing markets operated at a sub-regional level, normally across local authority boundaries, making the delivery of housing in one area dependent upon actions in other areas.

Proposals on HPDG were further contextualised in 2007 with the publication of a number of other consultation documents. The Planning White Paper (DCLG, 2007a) outlined the government's intentions within a wider framework of Local Area Agreements and a reduced set of national targets for local government generally, while the Housing Green Paper (DCLG, 2007c) made numerous references to the role of HPDG in helping deliver increased housing supply. However, little detail was included. Instead, it was left to a consultation paper on allocation mechanisms to answer many of the questions had been raised (DCLG, 2007d).

Improving performance through pricing

The shift away from linking resources to crude, process-related performance targets went wider than the HPDG. The Planning White Paper also asked for views on Planning Performance Agreements (PPAs) for major schemes (e.g. over 200 houses) that would not normally be expected to be agreed within the thirteen-week target. A PPA would be an upfront agreement between a developer and a local planning authority that would establish a timetable for delivering a decision. PPAs would then not be counted as part of the thirteen-week figures. A pilot project had shown that they could help deliver decisions more smoothly and with greater certainty for all concerned.

However, while accepting that PDG had led to perverse outcomes, the government was ready to use improvements in published development control performance to justify an increase in application fees. A consultation paper issued at the same time as the Planning White Paper set out proposals for increasing application fees:

> For the last five years, Planning Delivery Grant has partly bridged the gap between fee income and cost of service. Planning Delivery Grant brought significant improvements in service delivery but this grant régime is now in its final year in its current form. The issue of planning fees needs to be addressed soon if the reported improvements are to be maintained and further improvement delivered.
>
> *(DCLG, 2007e: paragraph 9)*

The DCLG estimated that the total cost of the development control service in England in 2006 was between £290 and 365 million per annum, and fee income for that year was only £232 million. To bridge this gap, the consultation paper proposed a fee increase averaging 23 per cent (DCLG, 2007e). A pilot project study of a premium service was also announced to allow local authorities to charge an additional premium fee of up to 20 per cent in return for a guaranteed decision within the eight to thirteen weeks. The possibility of locally determined fees was also discussed. Further changes were included in a companion consultation on the introduction of fees for appeals (DCLG, 2007f). Fees are not currently payable for appeals (though this may change), even though the system cost over £30 million per annum to administer in 2008/9. Like the mainstream planning system, however, the appeals system was under significant pressure. The Planning Inspectorate's caseload had risen from 14,000 appeals in 1997–8 to over 22,000 in 2005–6 and was forecast to rise to around 25,000 per year by 2010. While the Planning Inspectorate was also subject to targets, it was not achieving them – nor was it likely to do so with such projections of increases in its caseload.

Conclusions

PDG and the wider emphasis upon what Hood terms 'public service management by numbers' (2007: 95) is a phenomenon by no means unique to the UK or even England. Nevertheless, the scale, the significance of the centre in setting targets and monitoring performance, and the use of composite rankings of different targets linked to resources was unique and went beyond the approach found in other jurisdictions. Nor was planning singled out for this approach. Some of the issues undoubtedly arose because of the application of general performance management techniques to a sector that was subject to a range of sometimes competing objectives, including delivering high-quality, sustainable development and doing so quickly and cost effectively. Given the long history of attempts to speed up development control and the lessons that had emerged, it is mystifying that a more bespoke and suitable approach was not used – one founded upon an understanding of how the planning system separated plan from permission, on the degree of local discretion and on the need to emphasise outputs rather than processes. Actual improvements were disguised by the approach to performance measurement generally, which encouraged authorities to act strategically and led to perverse outcomes. The inevitable and predictable outcome was a mismatch between published performance and the experiences of applicants, as local authorities acted strategically to meet targets and secure funding. Yet performance *did* improve, and the profile of development control within authorities was enhanced to the benefit of planning and planners (HoC PAC, 2009: 3).

Nevertheless, the appetite for improvement and greater certainty within government and the development industry was insatiable: after five years of PDG there was still concern within government over development control and over planning constraints on housing supply and economic growth. As the British Property Federation put it to a House of Commons inquiry:

If you get more and more delays in the system then you are going to see developers who are less and less willing to undertake big and high risk schemes and that is going to become even more relevant given the current climate for development and the current attitude to risk ... planning oils the development system, and if planning is not working well, the development system is going to be severely impacted.

(HoC CLGC, 2008: paragraph 4)

Despite the recession and the collapse in housebuilding, the development industry was still trying to 'pin the blame' on planning for impeding delivery, largely ignoring the far more significant impacts around the reduction in demand for new development and the scarcity of finance. However, the relationship between developers and planning was itself changing as a result of PDG. One little-discussed further outcome of PDG was a culture change in development control following the emergence of a two-tier process as a consequence of performance-related targets. On the one hand are applications that conform to local policies and applicants that are prepared to 'toe the line'. On the other hand there are the 'non-conforming' applications, as indicated by one local authority chief planner: 'Actually, getting planning permission is really easy. You just apply for something that is in line with local policies and follow the guidance we give you about how to design layout and so on' (quoted in Allmendinger, 2009: 85). As another local authority senior planner put it: 'Refusals are made because developers have not done the work to demonstrate that their development will not cause harm' (ibid.).

There was also evidence in the same study that some local authorities and applicants/agents have made tacit arrangements to deal with applications in ways that minimise delays and streamline processes. Local authorities are keen to promote such relationships, and this would appear to favour knowledgeable agents in the process. There was a general attitude among the local authorities that the cost of delays and possible appeals to the applicants and not the local authority created an asymmetry of power. Those proposing a scheme were left with little choice but to agree to the requirements of a local authority. One chief planner stated that 'Delay occurs because developers want delay. They can modify the scheme or have an answer. We are only delaying it in the sense that the scheme wasn't the best scheme for the site' (quoted in Allmendinger, 2009: 85).

Other studies also echo these findings. The National Audit Office conducted its own investigation into performance targets and the use of the Planning Delivery Grant to incentivise local planning authorities. The report broadly backed up the picture presented in this chapter and came to the conclusion that, in terms of value for money, PDG had, at best, been 'mixed':

The Department has spent approximately £68 million a year on Planning Delivery Grant to increase the speed with which applications are handled. The combination of this grant and the setting of targets by the Department has succeeded in ensuring that Authorities give a higher priority to taking speedier decisions, and the proportion of major residential applications decided within the 13 week target has consequently almost doubled from 2002–03 to 2007–08. The Department, however, has no data on the average time taken to make these decisions and therefore on how it has changed over time. The Department's measure also does not identify whether there has been an improvement in the total time taken for schemes to progress through the development management process (from pre-application to the start of construction).

(NAO, 2008: paragraph 19)

In the light of the above and the findings of the Killian–Pretty review (Killan and Pretty, 2008), it is perhaps not surprising that the appetite for performance-related targets diminished within government and elsewhere. The Local Government White Paper *Strong and Prosperous Communities* (DCLG, 2006h) set out a commitment to streamline indicators and replace the Best Value regime with National Indicators (NIs) in 2008/9. NI157 sets out four development control performance indicators that relate to the eight-and thirteen-week determination targets. The government's target was that 80 per cent of major applications be determined within thirteen weeks by 2011. The near ubiquitous procedure and much criticised practice of refusing applications, encouraging withdrawals and resubmissions and using conditions to deal with matters rather than negotiating proposals is unlikely to change, given the advice that: 'Time spent in abeyance should be included in the total time taken (on no account should the clock be stopped) and the processing period must not be suspended awaiting amended plans nor restarted upon receipt of amended plans' (DCLG, 2008a: 18).

However, the incentives to act strategically in determining applications were diminishing through the HPDG scheme and the refocus from process to plan preparation and development delivery. Nevertheless, HPDG was not a panacea for the issues raised by PDG. Despite its failings, PDG had been clear and easy to administer. Local authorities had responded well and, despite the inevitable perverse outcomes and strategic behaviour, the process had been cost efficient. HPDG tried to achieve multiple objectives, some of which were beyond the control of the local planning authority. The main problem was one of control. Local authorities would be dependent upon the actions of others. McLean, Haubrich and Gutierrez-Romero highlight external constraints as a fundamental problem with performance measures: 'To be a valid measure, a performance score should be entirely uncorrelated with any background factor that a local authority can neither control nor be blamed or praised for' (2007: 116).

Government was simultaneously encouraging local authorities to work collaboratively in joint core strategies for their LDFs and even incentivising them to do so (see Chapter 5), while HPDG incentivised authorities to deliver housing in their own areas to receive income. There was a proportion of HPDG allocated to joint working authorities, though this amounted to 2.1 per cent in 2009–11 compared with 64 per cent of the grant allocated to housing delivery (DCLG, 2009c).

Finally, the threshold targets were likely to be too high to trigger payments, while the level of grant was itself unlikely to be sufficient to change behaviour. The allocation mechanism for HPDG sought to reward authorities that provide net additional housing completions above 0.75 per cent on a three-year rolling average (DCLG. 2009c: paragraph 9). This proportion was reduced to 0.65 per cent in 2009–10 (ibid.: paragraph 22). In 2007, UK housing completions amounted to a net addition of 0.86 per cent of stock (DCLG, 2009b) and fell to as low as 0.45 per cent net addition as delivery began to decline during the recession. In other words, even with a three-year rolling average, HPDG thresholds were likely to be too high for all but the highest growth areas.

Despite these changes there was a clear contrast with and tensions between this commodified approach to development control and that of spatial planning (Allmendinger, 2006). On the one hand, the government was pushing the notion of spatial planning and Local Area Agreements linked to a redefined notion of development control into development management (DCLG, 2010b), and other public bodies were encouraging more holistic and quality- focused benchmarks for planning (PAS, 2007, 2008). On the other hand, there remained a hard core of time-based performance indicators linked to financial incentives. The contrast and tension between the two remained unresolved. The more holistic approaches departed from (though embedded) traditional, measurable performance indicators in two main ways. First, the process emphasised peer review rather than external audit (though the Audit Commission could still play a role). The approach was also related more to what an area sought to achieve than to what government considered important and fitted in with the impact or output focused Comprehensive Area Assessment (CAA) regime.

The second main difference was the shift from development control to development management (DCLG, 2010b). While some are adamant that development management was not development control by another name (see PAS, 2008: 35) the performance targets still related to the time taken to determine applications. Despite the need for development control to be more proactive and positive, and the need to undertake it in a spirit of partnership and inclusiveness (DCLG, 2010b: 7), local authorities were still being asked to square the circle of speed and quality. This trade-off and tension is not unique to planning: 'New Labour target-setting and attempts to revive democracy have pulled in opposite directions over the past decade, with the former clearly ascendant' (Wilks-Heeg, 2009: 35). Development control was subject to what McLean et al. (2007: 115) term 'contradictory incentives' and 'output distortions'. The former occurs when organisations are pulled in two contradictory directions to achieve different targets – for example, spatial planning's emphasis upon coordination and integration across scales and the need to make quick decisions. The latter arises when attempts to achieve targets lead to significant costs on other performance – for example, refusing applications rather than negotiating. As McLean and his colleagues put it: 'when a measure becomes a target, it ceases to be a valid measure' (ibid.: 114).

7
DEVELOPMENT, INFRASTRUCTURE AND LAND TAXATION

Introduction

The government's growth agenda announced in *Sustainable Communities* (ODPM, 2003a) and reinforced through the Housing Green Paper (DCLG, 2007b) sought to deliver a step-change in housing and employment provision to 2020. Linked to this proposed growth were the thorny issues of infrastructure costs and development land tax, which resurfaced as they had so often done throughout the postwar history of planning. Attempts to capture some of the uplift in land value that results from planning permission has been a chronic theme of UK planning, and whenever it has been attempted it has largely ended in failure. New Labour's attempt was no different. The Housing Green Paper in 2007 reiterated Labour's commitment to increase house building to 240,000 units per annum by 2016, translating into 3 million new homes by 2020 (DCLG, 2007b). Such development would not be evenly spread but concentrated in the four 'growth areas' and twenty-nine 'growth points', the former involving 200,000 additional homes above previously planned levels by 2016 (ibid.: 24). With the impact of the credit crunch and recession on housing development from 2008, such targets became, at best, optimistic. However, the need to finance the infrastructure for such growth was a significant concern, particularly as the main existing mechanism for linking development to infrastructure – planning obligations secured through Section 106 of the Town and Country Planning Act 1990 – were widely criticised for being slow, costly and inefficient. Labour's initial solution was to propose in 2005 a form of development land tax named the Planning Gain Supplement (PGS) (HM Treasury, 2005a). PGS emerged as a recommendation from the Barker report, which argued for some form of flat-rate tax on the uplift in land values resulting from the securing of planning permission (Barker, 2004). PGS would tax this uplift and recycle a proportion of the money back to local areas to help pay for necessary development-related infrastructure. Existing Section 106 arrangements would be scaled back to focus solely upon site-related infrastructure, development mitigation and affordable housing provision. The government argued that PGS would be more transparent and certain than the existing system: 'The Government believes it is fair in principle for the wider community to share in the wealth created by planning decisions in their area, given the sizeable uplift in land value that planning decisions often confer' (HM Treasury, 2005b: paragraph 1.10). Tariff-based approaches, as an alternative to PGS, were rejected because they were: 'potentially limited in their scope, since they are most suited to addressing local concerns and may not easily deliver the investment required to support significant housing growth' (ibid.: paragraph 1.16).

After nearly three years of consideration and development, PGS was dropped in 2008 in favour of the Community Infrastructure Levy (CIL), a tariff-based approach that the Treasury had previously considered inappropriate. PGS managed to unite developers, landowners, local authorities and government agencies against it. Yet, there was also widespread support for reform of the existing system and an acceptance that development should make contributions towards necessary infrastructure. So why did the government pursue PGS and why, after three years of consideration and consultation, did it undertake an apparent U-turn and abandon PGS in favour of an approach that the majority of those in the industry suggested from the start? How could a government seeking such a significant increase in development activity propose a measure that was widely considered to be almost guaranteed to thwart development?

The previous four attempts to come up with a form of development taxation did not raise expectations that PGS would succeed. Many commentators and responders to the consultation papers pointed out that it was as flawed and doomed to failure as past attempts at land value taxation. But the reasons for the failure of PGS go much deeper and highlight many of the issues around Labour's approach to planning and policy, including the lack of a consistent approach and message within government over the function and objectives of planning. On the one hand, the ODPM and the Department of Communities and Local Government were concerned with planning as a tool to deliver sustainable communities, address worsening housing affordability and promote an 'urban renaissance'. However, others within government had a different view. One element of the Treasury saw planning as a 'burden on business', and this pointed to deregulation and the promotion of market-based alternatives. Another Treasury view focused upon the large uplift in land values that arose from the securing of planning permission and saw a way of capturing an element of this as a contribution towards necessary local infrastructure and general taxation. This tension – the left hand not knowing what the far left hand was doing – remained unresolved. The upshot was the promotion of a mechanism – Planning Gain Supplement – that would have achieved elements of the government's objectives but undermined the core aim of increasing housing supply. Perhaps as important in the downfall of PGS was the failure within government to understand the economics and dynamics of land and property markets, as well as repeating one of the core errors of the Thatcher administrations in adopting a policy that managed to alienate those charged with its implementation. Previous attempts to introduce some form of land tax had faltered, in part, on the reluctance of developers to proceed under conditions that would have reduced expected profits. Seeking increased housing supply at the same time as making development less profitable and potentially unviable strengthened the hand of those opposed to PGS. Why it took three years for this to be realised by some in government is less clear.

One of the ironies of the PGS episode was that the method of capturing some uplift in land value through Section 106 agreements was beginning to be more efficient and effective. The attitude towards planning obligations on the part of developers and landowners became more favourable and supportive, particularly in the light of PGS. As research showed, planning obligations were flexible, locally driven and would raise more than PGS in a range of circumstances (DCLG, 2006b; Knight Frank, 2006). They were also flexible enough to facilitate the standard tariff approach favoured by most in the development industry. The Milton Keynes 'roof tax', introduced in 2006, had demonstrated how a fixed development tariff could be introduced through the Section 6 mechanism. Ultimately, this approach would be the basis for the CIL. This chapter charts the rise and fall of PGS as an illustration of some key themes and lessons towards planning during the Labour era and more generally.

The Road to PGS

There have been four attempts by previous governments to secure for the public benefit a portion of the land value uplift resulting from planning permission: the 1947 Development Charge, the 1967 Betterment Levy, the Development Gains Tax introduced in 1973 and the Development Land Tax in 1976 (Cullingworth, 1980; Grant, 1999). While all have failed to a greater or lesser degree, the current system of planning obligations or planning gain under Section 106 of the Town and Country Planning Act 1990 has evolved over time to fulfil a similar role. Local authorities have increasingly used such agreements or obligations to pay for a range of community benefits, though they were originally envisaged as a means of mitigating development impact and allowing development to proceed through the provision of such essential off-site infrastructure as road improvements (Moore, 2005). Although successive governments sought to limit the more dubious uses of obligations, practice continued to evolve, and many developers and landowners legitimised the growth in the scope of obligations by entering into agreements. Criticisms of obligations have been aimed at the uncertainty of what will be required by authorities (Grant, 1999), the perception that they can be used to 'sell planning permissions' (Moore, 2005) and how they can be seen by local developers as a way of overcoming local resistance to a scheme (Bunnell, 1995).

A tension between the role of planning obligations as, on the one hand, a legitimate way of helping facilitate development and, on the other, delivering 'planning gain' for a community has never been satisfactorily resolved. However, against the backdrop of government proposals on growth, the subject rightly came to be considered by Kate Barker in her first review of planning. Building upon previous analyses of and proposals for reform of obligations/Section 106 mechanisms, the Barker interim report (Barker 2003: 142–61) echoed some of the earlier criticisms of obligations, as well as noting that planning obligations could be time consuming and could not offset some of the externalities related to development (e.g. they could not overcome some of the impacts of development upon communities). Obligations, the report argued, lacked transparency and were complex to coordinate, given the different demands from within local authorities (e.g. housing, conservation, etc.) and other public bodies such as health and police. Planning obligations could also, the interim report noted, reduce the residual value of land so as to make it unviable to develop. (One of the advantages of having a local, negotiated approach was to avoid such impacts, though there has long been a suspicion that some local planning authorities sought to thwart development through excessive planning obligation demands.) Finally, obligations did not always deliver the necessary infrastructure. Sub-regional infrastructure, including, for example, transport provision such as rail links, was identified as being necessary but largely absent from consideration and inclusion in obligations (Barker, 2003).

In reviewing other attempts at land taxation, Barker considered that the problems with previous approaches related to four main issues. First, there was the credibility of the tax. In many instances landowners withheld land for development, either in the hope that they could force government to repeal it when the supply of development land dried up or in the belief that the tax would be repealed in any case. Second, previous taxes were regarded as being too complex. Valuing land uplift for any tax was notoriously difficult and complex. Separating land value uplift as a result of planning permission from general land price inflation and accounting for the multiplicity of characteristics of land in any valuation required significant, costly and time-consuming effort, particularly in the absence of a market transaction. Third, previous taxes had been poorly targeted. As with other tax regimes, forward planning on the part of landowners and developers could reduce any liability. Finally, the rates at which some of the previous taxes had been levied neared 100 per cent of any uplift in land value, thus acting as a further disincentive for landowners. Nevertheless,

> These issues can be tackled ... by a better designed tax operating within a stronger development framework. For example, on the credibility issue, a more proactive use of compulsory purchase powers to acquire developable land, alongside a Development Gains Tax, might reduce landowner incentives to hold back land, given the possibility that it may be purchased compulsorily were it not to come forward for sale.
>
> *(Barker, 2004: 79)*

The Barker analysis of the existing system of obligations was driven by concerns around delivery and the perceived role of obligations in impeding housing development in particular (Barker, 2003). However, in the final report attention had shifted to include issues around equity and the costs of infrastructure delivery:

> The Government should actively pursue measures to share in ... windfall gains, which accrue to landowners, so that these increases in land values can benefit the community more widely. The value captured can be used as a funding stream for a number of other policies.
>
> *(Barker, 2004: 7)*

The notion that PGS would help delivery by supplanting obligations for off-site infrastructure was being accompanied by the acknowledgement that some form of taxation would itself reduce supply: 'In general, imposing a tax on an activity discourages its supply – but given the interaction of land supply with the planning system this effect could be expected to be small, provided that tax rate is not set at too high a level' (Barker, 2004: 8). For an inquiry that sought to facilitate an increase in housing supply, a recommendation that would decrease supply struck some as incongruous, to say the least. However, it was important, the report stressed, to see PGS as part of a package of measures that would increase housing supply. While the experiences of previous attempts to tax planning gain and economic theory led Barker to accept that PGS would lead to a decrease in housing land supply, this reduction would be more than offset *providing* other measures recommended in the report were either enacted or successful. Such interdependency in recommendations made a clear point: the government could not dine à la carte from the Barker recommendations.

Barker felt that a PGS system had three main advantages over other approaches.

- A tax based on uplift in land values at the point of the granting of planning permission would reflect local variations in land values, thereby minimising any disincentive to landowners to bring land forward.
- Varying tax rates could be used to help enforce other policy objectives, such as promoting brownfield over greenfield development.
- In locations that included Growth Areas, tax rates could also be varied to address weak housing demand.

However, the report went on to note that the success of any PGS would be dependent upon two factors. First, credibility was critical. Landowners and developers must feel that the system was fair and that it was to be permanent. Second, some transitional arrangements would be required. The assumption behind PGS was that the tax would fall largely on landowners and be reflected in reduced land prices, with little impact on developer profit and house prices. However, developers drew attention to the fact that they already had land banks, which had been assembled under different assumptions of land cost and development viability. They also pointed out that margins on certain types of development, such as brownfield, were less in any case, given the higher costs associated with land assembly, decontamination, etc.

The Barker recommendations envisaged that the introduction of PGS would be accompanied by the scaling back of the scope of Section 106 agreements or planning obligations to site-related mitigation measures and the provision of affordable housing. For some, this mixed approach of PGS and Section 106 was a surprise, given the strong criticisms in both the interim and final Barker reports of the use of Section 106 mechanisms. The final report argued that:

- the value of contributions achieved varies considerably between areas, and even between sites, in the same housing market locality;
- negotiations can take many months, occasionally years, and are costly in both local authority and developer time and resources;
- there may be asymmetries in negotiating expertise between the two parties, leading to unsatisfactory outcomes;
- local authorities are not always aware of the level of planning contributions that might reasonably be expected in a given development, due to the non-transparent nature of the system; and
- some local authorities may misuse Section 106 to delay or discourage development, by asking for unreasonably onerous levels of developer contributions.

(Barker, 2004: paragraph 3.48)

These issues would remain under the proposed approach, which retained the Section 106 element for on-site matters, though they would arguably be leavened by the more restricted form of obligations. The last significant recommendation was that the PGS levy could be varied so as not to discourage brownfield development or areas where the market was weaker – i.e. the rate could be used to incentivise certain forms of development. Finally, a 'proportion' of the of the resulting revenues should be transferred to local authorities, be at least as much as would have been raised through obligations and be used as the authority 'thought fit' (Barker, 2004).

The government welcomed the Barker recommendations (HM Treasury, 2005a) and issued a consultation paper in December 2005 (HM Treasury, 2005b) taking on board Barker's suggested term of Planning Gain Supplement (PGS). In the government's view PGS would:

- be an essentially local measure and, as such, a significant majority of PGS revenues would be recycled to the local level for local priorities. This would help local communities to share better the benefits of growth and manage its impacts, as well as ensuring that local government overall would receive more funding than was raised through the Section 106 approach;
- dedicate additional investment in the local and strategic infrastructure necessary to support growth. The government anticipated that an 'overwhelming

majority' of PGS funds would be recycled within the region from which they derived; and
- fund strategic, regional and local infrastructure to ensure growth is supported by infrastructure in a timely and predictable way. Local and regional stakeholders, including business, would play an important part in determining strategic infrastructure priorities to help unlock development land (HM Treasury, 2005a: paragraph 3.14).

Attention in the development and planning communities immediately focused on a number of issues, among them the recycling of a 'significant majority of PGS revenues', the aim of increasing the overall amount of money captured through PGS than had been raised through Section 106 arrangements, and the objective of using some of the money to provide regional as well as local infrastructure. The consultation document went further than Barker in discussing some of the detail on how PGS would work.

- The uplift in land value would be calculated immediately after planning permission had been granted.
- The basis of calculating values would be on the assumption of an unencumbered freehold regardless of actual ownership.
- A new Development Commencement Notice would be introduced. This notice would be issued by the developer on commencement and be used to identify tax liability.
- PGS would be excluded from home improvements but would fall on commercial developments.
- Planning obligations would be scaled back along the lines suggested by Barker.
- Infrastructure provision paid for through PGS would need to be 'timely and predictable'.

However, there were also some significant differences between the Barker model and that consulted upon by the Treasury – specifically, local authorities would not be able to use PGS money as 'they saw fit', and the government was considering whether a lower PGS rate would apply to brownfield development as originally recommended (HM Treasury, 2005b).

One of the main problems in responding to the consultation paper related to uncertainty over the indications of how much the government intended to raise through PGS. This target would be essential in setting the rate and this, in turn, would affect development viability. The only hint given in the consultation document was that PGS would raise more than had been brought in under existing Section 106 arrangements. But as the latter amount varied greatly, this added little to the understanding of how PGS would operate. The government refused to respond to calls for some indication of the rate on the grounds that a Comprehensive Spending Review (CSR) in 2007 would provide an assessment of infrastructure costs in the Growth Areas and how much would be required through

PGS to pay for this. Such uncertainty only helped fuel the debate over whether PGS was a tax or a tariff. The consultation paper did not help clarify this ambiguity and, given the vagueness over the rates to be applied and the proportion to be recycled, the feeling was that PGS was to include a large element of the former.

Apart from the Royal Town Planning Institute (RTPI) and Town and Country Planning Association (TCPA), which both welcomed the redistributive element of PGS and saw few problems or issues, reaction to the proposals was largely hostile. Despite the strong arguments in the Barker report, both the RTPI and TCPA saw no case for having differential PGS rates for brownfield and greenfield developments and for different areas. While no levy figure was included in the consultation, it was widely being discussed as being 20 per cent of the uplift value once planning permission had been granted. The argument against differential rates was that, since the uplift was smaller on brownfield sites or areas with weaker demand, the PGS would be consequently smaller. The actual processes of land and property development pointed to a less simplistic picture. In evidence to the House of Commons Communities and Local Government Select Committee, both the Royal Institution of Chartered Surveyors (RICS) and the British Property Federation (BPF) argued strongly that undifferentiated PGS rates would undermine wider government objectives around promoting brownfield development. Contrary to the government's view, the property industry put forward several arguments in evidence to the committee in the House of Commons.

- Brownfield land tends not to be bought in a linear or sequential way but is usually assembled over long periods, using assumptions about future value. PGS would shift the balance of value for some parcels bought in earlier periods when no PGS was envisaged and create uncertainty over appraisals of future development value.
- Many issues that affect land value in brownfield locations, for example, contamination, are not identified until development occurs. PGS would be calculated and paid before development commences.
- Essential off-site infrastructure required to make brownfield schemes viable or even developable would be shifted from the planning obligations regime and replaced by a system that required the public sector, usually a local authority, to receive money from PGS via government and undertake the developments. This would create uncertainty over when and to what extent necessary infrastructure development would occur.
- Many brownfield schemes are mixed use, as encouraged by government. Some aspects would be exempt from PGS and other aspects would be liable, distorting the market and making some forms of development more attractive than others.
- The government and Barker had assumed that all land would be vacant and freehold possession, while, in reality, interest may be freehold and involve a range of existing buildings with different use and value potential. Valuing these complex situations for tax purposes at the point of planning permission

involves making numerous assumptions concerning possible future uses and ownership.
- Developers often modify planning permissions or hold more than one permission on a site. Requiring payment on the grant of permission for different schemes rather than actual development may require multiple PGS liabilities.

(HoC CLGC, 2006)

The Royal Institution of Chartered Surveyors thought that 'there is a strong risk that, in many regions, it [PGS] could curtail development of all kinds' (RICS, 2006: 1). Work for the South East England Regional Assembly (SEERA, 2006) estimated that PGS would raise an additional £3 billion (assuming a 20 per cent rate) over and above that expected to be raised through Section 106 arrangements over the twenty-year period of the Regional Spatial Strategy. Nevertheless, as there was so much ambiguity in the proposals, combined with a lack of clarity over what proportion of the extra money raised would be recycled, SEERA could give PGS only a guarded welcome. Other work by Crook and Rowley (2006, 2007) was less supportive and found that the rate would have to be set at 25 per cent to achieve revenues currently raised by obligations in the South-East, let alone to raise 'significant additional funds' – hardly the 'modest' amount discussed by the government. Research commissioned by the British Property Federation, Home Builders Federation, CBI and RICS also highlighted how PGS would not bring in as much as the current approach to obligations unless set at an immodest rate:

> The total planning gain contribution of all eighteen case studies under the current Section 106 system is approximately £375m. Under a scaled back Section 106 system, together with PGS, planning contributions would be approximately £195m for a PGS rate of 10 per cent, £279m for a rate of 20 per cent, and £363m for a PGS rate of 30 per cent. Overall, for this selection of case studies, this represents a reduction in planning gain, in relation to the case study examples included in the research, of 48 per cent for a PGS rate of 10 per cent, 26 per cent for a PGS rate of 20 per cent, and 3 per cent for a PGS rate of 30 per cent.
>
> *(Knight Frank, 2006: 4)*

The lack of agreement on the impact of PGS reflected the wide range of assumptions that had to be made in coming up with such figures. Such assumptions were themselves a function of the ambiguity in the proposals and helped undermine the government's argument that PGS would create greater certainty over infrastructure financing and increase the amount of money available.

The issues raised by those in the property industry and the lack of certainty over the proposals were compounded by the lack of a consistent message from within government. Around the same time that the Treasury-led proposals on PGS were being consulted upon, the Department of Communities and Local Government issued *Planning Obligations: Best Practice* (DCLG, 2006a) and commissioned research to demonstrate how planning obligations were being used in practice in order to support a 'more informed debate' (DCLG, 2006b: 6). The latter work was ambiguously entitled *Valuing Planning Obligations*. The Audit Commission also published its own research into how local authorities could best secure developer contributions through Section 106 agreements (Audit Commission, 2006a).

While acknowledging that there was a possibility of PGS being introduced no earlier than 2008, the *Best Practice* guide nevertheless tried to address some of the criticisms that had been aimed at Section 106 mechanisms by Barker and others (DCLG, 2006a). The new system of development planning and Development Plan Documents (DPDs) introduced in 2004 was recommended as the most appropriate mechanism by which local authorities could set out in advance what they would expect in terms of contributions through the current Section 106-based regime. Such an approach would increase certainty and transparency while allowing flexibility to adapt to individual site characteristics. Coincidently, previous critics of the Section 106 mechanisms also began to consider the alternatives. Liz Peace, chief executive of the British Property Federation, in her evidence to the House of Commons Communities and Local Government Committee on 24 April 2006, said:

> it is true the industry has complained extensively about the way section 106s are handled, but there is a huge amount of divergence between local authorities who handle it well and local authorities who handle it badly and a great raft of local authorities who do not do it at all . . ..

The DCLG's *Valuing Planning Obligations* document backed up the widely held impression that the number, scope and value of obligations had increased significantly in recent years, as had the use of standard charging. Standard charging or 'roof tax' approaches set out in advance either a single amount per dwelling to be paid to cover all off-site infrastructure and some mitigation or a range of up-front charges that the developer could expect. Over 60 per cent of authorities used standard charging for certain types of obligation, such as open space provision, and, in the majority of cases, raised more through standard charging than through negotiated obligations (DCLG, 2006b: 20). In other words, practice was evolving towards the kind of transparent and up-front approach advocated by Barker, while the amounts being secured were delivering a share in the land value uplift. The (then) ODPM had encouraged authorities to move towards standard charging and had commissioned research on the use and possible future of standard charges as the Barker inquiry had got under way (ODPM, 2004a).

The confluence of a tax-raising and infrastructure-funding mechanism, combined with vagueness around the rate at which PGS would be set, if it would be hypothecated, how much would be recycled to the local level and which public bodies would be charged with spending the money, meant that it was difficult to argue against PGS. The government was caught between needing to provide more detail and not being clear itself whether PGS was aimed purely at development infrastructure or whether it would have an additional tax-raising objective. As a consequence, local authorities were largely lukewarm to the notion: while largely welcoming the possibility of more resources, they were unsure about how much of the money would find its way back to them. The development industry similarly had grown used to negotiated planning gain through Section 106 agreements. PGS introduced uncertainty about future rates and additional costs in migrating to a new approach while not adding any advantages over the current system. Indeed, there was considerable uncertainty over how PGS would be calculated and built into development appraisals. It is not surprising, therefore, that few came out unequivocally in favour of the mechanism.

Nevertheless, the government's summary of responses to its consultation paper concluded that there was general support for some form of development tax on land value uplift that results from the granting of planning permission, though it highlighted a wide range of concerns that had arisen from those consulted (DCLG, 2006c). This conclusion was broadly supported by a House of Commons Communities and Local Government Committee (HoC CLGC 2006). The principle of securing a proportion of the uplift in land value to pay for infrastructure would help meet the nation's requirements for increased housing supply. However, the committee argued that PGS needed to be implemented as part of a package of measures and not be used as a means of replacing funds. In other words, it endorsed the notion that PGS would bring in funds over and above those raised through planning obligations but that it would not be a general tax-raising mechanism.

Continued questions and criticisms throughout 2006 culminated in an announcement as part of the pre-budget report on 6 December 2006 and the publication of three further consultation papers. In response to the reservations and criticisms concerning the vagueness of the approach, the Treasury stated that the government 'will move forward with the implementation of PGS if, after further consultation, it continues to be deemed workable and effective' (HM Treasury 2006a: 1). The growing possibility that PGS might be defendable in principle though be unworkable in practice provided the backdrop to the launch of three further consultation documents that added some further detail to how the government envisaged the system working – specifically, that:

- PGS would be levied at a 'modest' rate and would apply to residential and non-residential development;
- at least 70 per cent of PGS revenues would be recycled back to the local authority area from which the revenues derived for infrastructure priorities, with the remainder being returned to the regions to help finance strategic infrastructure projects; and
- a workable and effective PGS would not be introduced before 2009.

The three consultation papers reflected three of the four broad issues that had concerned those consulted: how PGS would be valued (HM Treasury, 2006b), how it would be paid (HM Treasury, 2006c) and how the existing system of planning obligations would be reformed (DCLG, 2006d). Questions concerning how the money would be spent and recycled remained unanswered.

The attempt to explain better how PGS might work in practice led to more criticisms and objections than clarification. The Treasury's view that it would only be implemented 'if workable' acted as an invitation to those opposed to it to further their efforts. The *Valuing Planning Gain* consultation paper (HM Treasury, 2006b) set out the government's desire to keep PGS valuation requirements as simple and straightforward as possible and to avoid the complexities faced by developers of earlier development gain taxes. And yet, as many pointed out, the proposed process was fraught with uncertainty and ambiguity. PGS valuations would be undertaken on a self-assessment basis, with developers using valuers to assess both current use values and planning values to calculate any uplift. What to include and exclude in valuations, how to undertake them and how and in what circumstances the Valuation Office and Her Majesty's Revenue and Customs (HMRC) would challenge developer's valuations were important though absent issues. As English Partnerships pointed out: 'under self-assessment there would be the potential for disputes to arise between the developer and HMRC regarding the amount of the liability, leading to delays in raising revenue' (English Partnerships, 2006: 7).

Paying PGS (HM Treasury, 2006c) sought to provide further information on how developers would go about applying for a start notice and pay HMRC and how and in what circumstances HMRC and the Valuation Office would intervene if they felt that self-assessment was inaccurate. Concern was expressed that, far from being the simple and straightforward approach the government envisaged, PGS would add to local authority workloads through information requests for valuation purposes. Valuers would need to know about previous applications, adjacent proposals and permissions, and a host of other information in order to come up with a value. Valuations based upon such information would include assumptions all of which could be challenged by the Valuation Office. In addition to this workload, local authorities would still have to negotiate over planning obligations for on-site mitigation measures and affordable housing. There was, therefore, scope for disagreement and protracted negotiations between the developer's valuer and the Valuation Office. A further issue concerned the proposal to make subsequent owners liable for PGS if not paid by the developer. Buyers' solicitors would need to be sure that payment had been made and that their clients would not be liable for any outstanding monies that might arise from disputes or appeals when land and property is sold on. This would imply a time limit on challenges to self-assessments and valuations by the Valuation Office. It would therefore be rational for developers and their valuers to submit lower uplift appraisals in the hope that there would not be a challenge from the Valuation Office. Developers and investors would also have to factor possible challenges into their evaluations, creating uncertainty in appraisals. As such appraisals often underpin land values, developers would be in danger of paying too much for land.

The final consultation document concerned how planning obligations would be revised and scaled back under a PGS regime (DCLG, 2006d). Again, as with the other two consultation papers, the attempt to provide clarity actually created greater uncertainty. Developers had become used to 'obligations creep', as local authorities had sought to extend the scope and value of planning gain to maximise revenues. While some of the ways in which this had been achieved were not in the spirit of government advice, developers had largely gone along with it as the 'price to pay' for planning permission. The alternative would be a drawn-out and costly appeal, at the end of which there was no guarantee of success. In other words, developers were happy to pay a premium for certainty.

What *Changes to Planning Obligations* set out was an approach that sought to limit obligations post-PGS to the mitigation on the site itself and any proportion of affordable housing:

> If the requirement related directly to the viability of the *physical environment of the development site or the need for a proportion of housing to be affordable*, it could continue to be the subject of a planning obligation. If, on the other hand, the requirement related to the site's social or community infrastructure, it would no longer be included within scope.
>
> *(DCLG, 2006d: paragraph 24; emphasis added)*

The consultation paper went on to stress that off-site matters could still fall within planning obligations providing that they had an 'appropriate relationship to the physical environment of the site' (ibid.: paragraph 26). To many, this ambiguity was a licence to local authorities to push the boundaries and argue for an enlarged scope of obligations; developers had the choice of entering into this obligations or appealing against a refusal of planning permission. The underlying pressures on local authority finances that led to the expansion of obligations had not gone away. The fear was that developers would pay PGS and be expected to enter into costly obligations despite proposed limits.

The overall result of the three papers was to strengthen the resolve of those opposed to PGS. The CBI had been leading calls for it to be abandoned and summed it up thus:

> The Chancellor stated that PGS would not be pursued if found to be ineffective and unworkable. Officers at HMT have since commented that they do not foresee any 'show-stoppers' to PGS. The CBI in contrast would assert that the cumulative impact of PGS on the UK economy as evidenced in this response should be seen by Government as a 'show-stopper'. In our view PGS remains unworkable and ineffective and we believe this policy should therefore be dropped.
>
> *(CBI, 2007: paragraph 29)*

A way forward?

Despite opposition, there was widespread understanding on the part of the development industry of the need to focus on infrastructure if the Barker and government targets on housing supply were to be met. As the Home Builders Federation put it: 'it is impossible ... to contemplate increasing house building rates from 150,000 to 200,000 homes a year without putting in place a properly thought through infrastructure that will adequately service the needs of the people who live in those homes' (HoC CLGC, 2006: 6). Yet the government seemed committed to PGS in the face of almost overwhelming opinion against it based upon the widespread feeling that it would hinder delivery. Four issues stood out.

1   *Credibility*   The government had been sending out mixed signals on PGS and the future of obligations. On the one hand, the DCLG had been pushing a reformed approach to obligations while simultaneously acknowledging that PGS might be introduced. The result was that those opposed to the system were pushing at an open door on its credibility. A coordinated campaign emerged, with the British Property Federation, the Home Builders Federation, London First and the Major Developers Group commissioning research and conducting a campaign questioning PGS in the pages of the industry press. Support from local authorities and other public bodies was, at best, lukewarm, with a range of questions and issues being highlighted. With a united industry campaign, mixed messages from government, and alternatives in the shape of the increasing use of standard charges and the known approach of negotiated obligations, it was not difficult to predict the outcome. The other factor that must have been on the minds of those opposed to PGS was that the previous attempts at a form of development land tax were introduced by Labour governments, only to be repealed by incoming Conservative governments. Thus, like previous attempts at development land taxation, PGS lacked credibility.
2   *Complexity*   It is a cliché to say that valuation is an art not a science. PGS would require developers to submit self-assessments and enter into negotiations with the Valuation Office over the assumptions and underlying valuation methods. To reduce the risk that they had overvalued sites, developers would need to undertake far more detailed, costly and time-consuming site investigations before development began. Advice from the government did not help in addressing such complexity:

> When a valuation is complex and subject to large tolerances then this will be recognised by valuers in the Valuation Office Agency (VOA) and Valuation and Lands Agency (VLA) when checking valuations on behalf of HMRC. If the self-assessed valuation submitted by the taxpayer is pitched at a reasonable level within the range then the VOA/VLA would not normally seek to challenge the valuation.
>
> *(HM Treasury, 2006b: paragraph 8.25)*

As some respondents pointed out, complex valuations tended to arise in exactly the kind of developments such as brownfield that the government was seeking to encourage. What was regarded as 'reasonable' or 'complex' was, of course, open to widespread interpretation. As some pointed out, the PGS process and its inherent uncertainty actually added to the complexity of most valuations.

3   *Targeting*   Although Barker had recommended a variable rate PGS that would be sensitive to different markets and would help reinforce wider objectives around making the maximum use of previously developed land, the government, backed by the RTPI, had argued for a flat rate. The rationale was that PGS would only be levied on uplifts in value and that, as had they smaller uplifts, this would not disadvantage brownfield developments. This was in the face of widespread contrary views:

> PGS as a tool for deriving a proportion of development value may work satisfactorily for small scale development on greenfield land, in areas where large increases in land value arise with the grant of planning permission, as indeed the examples used in the Government consultation documents illustrate. Where the development process is more complex, the introduction of PGS could lead to severe delays and increased development risk due to problems in negotiating and agreeing the appropriate PGS liability.
>
> *(BPF, 2007: 4)*

Brownfield developments were more complex, time consuming and costly. Adding to that complexity would act as a further disincentive to developers, regardless of the actual amount of PGS.

4   *Tax rates*   The government did not offer a possible PGS rate and instead talked of a 'modest' rate. In this vacuum a figure of 20 per cent emerged (though it is unclear where this came from). Rates for previous forms of development land tax had been much higher, but the fact that the rate could be changed worried many, especially as land assembly and development processes could take years. Some land parcels could have been assembled under different assumed PGS rates, affecting viability, or, if banked for some time, could have been bought under the assumption of no PGS. Equally, because of uncertainty over the rate, work on how much PGS would raise came to widely different conclusions, and this in turn would affect the infrastructure that could be delivered. Developers also worried that there would be upward pressure on rates from local authorities and others, and that this would lead to landowners withholding land until the rate was reduced or PGS was abolished. While such variations in values and the assumptions behind costs were part and parcel of the risks taken by developers, the uncertainty created was felt to discourage development on all but the most straightforward sites.

The more the government tried to clarify and develop the concept of PGS from 2005 onwards, the more difficulties became apparent. While it was clear to most (though not government) that PGS did not meet the government's own tests, it was equally clear that there was a workable and existing alternative. Variously labelled a 'planning tariff' or 'roof tax', it was an approach that underpinned the large planned growth in Milton Keynes, though was also increasingly being used in a wide range of local authorities (DCLG, 2006b). Under such an approach, estimates are made of required infrastructure to support development over a plan period. Existing planned public and private spending are factored in, and the resultant figure is divided among the planned development as either a standard charge per house or, in the case of commercial development, for floor area. In Milton Keynes the amount came to £18,500 per house, with a total income of £300 million over ten years, including social housing. However, the total cost of infrastructure related to planned growth (taking in improvements to the M1 motorway) was in the region of £1.6 billion, so the tariff covered only a small proportion of costs. Other places, such as Reading and Cambridge, produced lists of standard or typical contributions to help provide greater certainty for developers. All approaches are based on Section 106 agreements with a 'standard tariff' – in effect, the 'front end'.

However, the 2004 Barker report followed extensive consultations in 2001 (DTLR, 2001c) on a mandatory tariff and in 2003 (ODPM, 2003b) on an Optional Planning Charge, neither of which received support. Thus, the government in the guise of the DCLG was not disposed to tariffs *per se* while the Treasury-led Barker review had come down in favour of PGS. As the debate over how best to proceed ensued, the ODPM and then the DCLG began to reform obligations by issuing new guidance. Either as a result or because of experience, many local authorities began to improve their obligations processes by, for example, introducing a form of standard tariff as supplementary planning guidance and negotiating obligations in parallel with rather than sequentially after planning permission. These and other changes increased certainty and reduced delays while raising significant amounts for local infrastructure and development impact mitigation (DCLG, 2006b). Authorities were voting with their feet and developers broadly backed this approach, particularly as the spectre of PGS loomed. More local authorities would have moved towards such an approach if it were not for PGS: the 'threat' of PGS is now one of the major blockages in agreeing new tariff schemes' (Walker, 2007: 7).

The 'Milton Keynes approach' provided a very useful alternative for those opposed to PGS, even though the government rightly pointed out that the circumstances there were unique, through the central role of (the then) English Partnerships in loaning the money for necessary infrastructure and the tariff effectively paying it back. This left the government in the position of attempting to scale back obligations, even though a form of system had begun to evolve which overcame at least some of the problems associated with them. At the same time it was increasingly clear that PGS would have virtually no support from those charged with its implementation and could actually decrease housing land supply and development.

The summer of 2007 saw three events that sealed the fate of PGS. The first was the change of prime minister in June 2007 and the subsequent series of speeches and announcements by Gordon Brown that sought to define his objectives as leader. Among these were tackling worsening housing affordability and increasing housing supply. The second was the issuing of the Housing Green Paper *Homes for the Future: More Affordable, More Sustainable* (DCLG, 2007b). The new targets for housebuilding and supply to 2020 were ambitious and, along with the provisions in the Planning White Paper published earlier in the year (though under a different prime minister) (DCLG, 2007a), included a raft of planning-related proposals, such as a new system for the delivery of major infrastructure and using planning to help improve economic competitiveness. The proposals on housing supply and development were clear – planning had to deliver. The potential risk that PGS would reduce housing supply, as argued by those opposed to it, ran contrary to this. Finally, the government had come in for significant criticism over its introduction of Housing Information Packs (HIPs). HIPs were introduced in 2004 and required sellers of properties to provide information to potential buyers. Criticism of the cost and usefulness of HIPs had been particularly fierce, and the Royal Institution of Chartered Surveyors had successfully challenged their introduction through a judicial review in May 2007. The government and the ministers involved were starting to appear (and were being portrayed in the media) as out of touch with housing and development issues.

The outcome was that the ground began to be prepared to back-pedal on PGS. The Planning Green Paper conceded that it was not popular for a variety of reasons and offered a further round of consultation: 'The Government, in the light of the representations it has received, is offering local authorities and developers a further opportunity to discuss alternative approaches to PGS' (DCLG, 2007a: paragraph 33). Four alternatives were offered to facilitate discussion:

> *Option A*   This would involve a reduced PGS rate (a legitimate question would be 'reduced from what?', as no rate had been advanced by government) though with no changes to the current scope of obligations. PGS revenues would still be recycled to the region from which they were raised.
>
> *Option B*   PGS would be limited to greenfield sites only. As greenfield sites generally experience a higher uplift value from planning permission, PGS would seek to capture some of this while helping to improve the viability and attractiveness of brownfield sites. Planning obligations would remain for other sites.
>
> *Option C*   An expanded system of obligations would allow local authorities to move towards standard charges and widen the scope of obligations to include strategic and sub-regional infrastructure.
>
> *Option D*   This was a standard planning charge based on the Milton Keynes approach, which relied upon Section 106 agreements. The government

would legislate to formalise standard charges. The ability of authorities to choose either obligations or standard charges was not discussed.

The Green Paper stated that PGS remained the preferred approach and set a deadline for responses to the four options of 15 October 2007. Reactions in the professional press presented the consultation as a U-turn forced by a concerted campaign led by the British Property Federation and others. To nobody's surprise, the outcome was the rejection of PGS:

> Following discussions with key stakeholders, the Government will legislate in the Planning Reform Bill to empower Local Planning Authorities in England to apply new planning charges to new development, alongside negotiated contributions for site-specific matters. Charge income will be used entirely to fund the infrastructure identified through the development plan process. Charges should include contributions towards the costs of infrastructure of regional or sub-regional importance. Legislation implementing PGS will therefore not be introduced in the next Parliamentary session.
>
> *(HM Treasury, 2007b: paragraph 6.17)*

The new approach, enabled in the Planning Act 2008, was termed the Community Infrastructure Levy (CIL). CIL is a standard charge based approach for off-site infrastructure:

> The Community Infrastructure Levy (CIL) will be a new charge which local authorities in England and Wales will be empowered, but not required, to charge on most types of new development in their area. CIL charges will be based on simple formulae which relate the size of the charge to the size and character of the development paying it. The proceeds of the levy will be spent on local and sub-regional infrastructure to support the development of the area.
>
> *(DCLG 2009e: paragraph 1.1)*

CIL was greeted with relief by most, particularly as local authorities could 'opt in' to it. However, authorities that had introduced tariff approaches based upon existing Section 106 powers would be expected to migrate towards CIL once it became operational in April 2010. In addition, as the scope of Section 106 would be reduced to site-related matters and affordable housing, there was motivation for those authorities that sought off-site infrastructure and mitigation contributions to take up CIL. Unlike with PGS, a variable rate (including a zero rate where appropriate) could be levied. As Lord (2009) points out, CIL was envisaged as a partnership-based approach, underpinned by Local Infrastructure Programmes, financed by mutually agreed Local Infrastructure Funds and delivered by public–private bodies and administering vehicles, Local Infrastructure Groups. However, this partnership model began to look unlikely in the face of reduced development activity during the recession and the need to renegotiate many existing Section 106 agreements as land values fell. The ability of CIL to accommodate such changes was questioned and the advantage of certainty began to look less important than flexibility. One outcome was little enthusiasm around CIL. Research by Drivers Jonas of 363 local authorities (93 per cent response) in January 2009 found that:

- 59 per cent of the authorities interviewed said they had no current plans to introduce a CIL;
- only 22 per cent of the authorities suggested that they would be implementing a CIL either now or in the future. For some, the availability of money to implement it is a key factor. Consequently at least 10 per cent (37 authorities) say they will definitely not be implementing a CIL;
- there is widespread confusion following the earlier changes in Planning Policy Statement 12 (PPS12) on how the CIL may be incorporated into the policy frameworks, particularly now that many authorities are at more advanced stages of the LDF process; and
- despite extensive Government support and consultation at least 1 per cent of the authorities interviewed had not heard of CIL or how it could benefit the local area.

(Drivers Jonas Deloitte, 2010)

The above figures contrast starkly with the DCLG's own estimates of between 65 per cent and 78 per cent of authorities taking up CIL. Based upon such estimates, the department predicted CIL would increase contributions nationally to between £4.1 and £6 billion above money raised through obligations (DCLG, 2010a).

Some authorities are moving ahead with tariffs and preparing the ground for the migration to CIL. Cambridgeshire County Council, Cambridge City Council and South Cambridgeshire Council, for example, see it as having a number of advantages over negotiated planning obligations, including a stronger legal basis and the ability to capture contributions from smaller schemes that would otherwise have been ignored (Cambridgeshire Horizons, 2009). It is no coincidence that the lead body in the Cambridgeshire scheme, Cambridgeshire Horizons, is a sub-regional growth delivery vehicle and that CIL is aimed, in part, at funding sub-regional infrastructure (Chapter 5). The Cambridgeshire scheme also benefits from being in an area of high demand. Other authorities, particularly those in which underlying demand is weaker, have far less scope for capturing contributions. For such authorities, the scaled-back, on-site-related Section 106 mechanism is likely to be more appropriate.

Conclusions

The Planning Gain Supplement highlighted a range of issues around the New Labour era and a wider misunderstanding of planning and development issues.

PGS was integral to the growth agenda and the need to link development to necessary infrastructure in a more strategic and transparent way. The principle of development having to contribute to necessary infrastructure was accepted by all. Yet some senior figures in the development and planning industry described the idea of PGS as 'barking mad' (*Planning*, 9 December 2005) and 'curious' (Lord, 2009: 335). The credit crunch and recession in the late 2000s put much development on hold and led to developers seeking to renegotiate previously agreed obligations. The downturn in development activity has raised a number of further questions around the relationship between development and infrastructure provision, particularly on the need for flexibility, which is not a key feature of either PGS or CIL. It is likely that the impact of the recession on development activity and economics will further entrench planning obligations as the tool of choice by developers and local authorities. In relation to the focus of this book, three issues stand out.

Appropriate mechanisms

One of the outcomes of the PGS episode was the widespread relief that it would not be introduced, along with a nostalgic feeling that the current system of obligations, while not perfect, broadly worked:

> While the CBI has been critical of planning obligations this has been more about the implementation of section 106 than the principle. While we would welcome further improvements to section 106 negotiations we do not see this as reason enough to introduce PGS.
>
> *(CBI, 2007: paragraph 25)*

The possibility of PGS led to some reflection upon the positive elements of obligations. Two points rose from this. First, there was concern over how PGS would break the link between a community and new development:

> Many developers are also long-term investors in communities and want the freedom to be able to negotiate the appropriate mitigation to impacts arising from the development which the current system of section 106 allows for. The proposed system will discourage community engagement by encouraging developers to focus solely on their development-site environment and expect everything else to be delivered via their PGS contribution.
>
> *(BPF, 2007: 5)*

Second, there was an issue of control over the delivery and timing of off-site infrastructure provision:

> The current Section 106 arrangement, although contested and negotiated by developers and planning authorities, normally results in a planning gain contract that is viable to the developer whilst meeting wider community

objectives. This approach is acceptable to developers on the basis that they retain some control over the delivery of the community benefits, since these will add value to the new development that is being undertaken. Under the PGS arrangement, this control would be lost as community benefits related to a site could no longer be negotiated under Section 106 agreements.

(Knight Frank, 2006: 5)

Both of these issues remain extant under CIL. There was also a growing concern within the property and development industry over the lack of flexibility in the CIL mechanism to vary the charging schedule if economic conditions change, with some claiming that this would take around two years (*Planning*, 2 October 2009). Another concern around flexibility arose over the ability to negotiate site-specific charges to reflect the specifics of a particular development. The fear was that either the charge would be set too high, and therefore make a scheme unviable, or be set too low and so not be able to fund the necessary infrastructure. The possibility of having an exceptions policy to allow a local authority to negotiate over a charge for a specific site was rejected by government. While CIL may be an appropriate mechanism in some rather specific circumstances, the 'one size fits all' attitude of government runs against the desire of some areas to adopt such tools to fit their own circumstances. It also flies in the face of the natural evolution of planning obligations towards a tariff-based approach in some areas.

Confused objectives

An issue that was never satisfactorily addressed was the difference between PGS as a general tax-raising mechanism and the use of PGS funds to pay for infrastructure. Barker had weighed the alternatives and carefully crafted an approach that sought to achieve a range of objectives, with an eye on delivery and infrastructure. However, the government ignored elements of her proposals, particularly the need for a variable rate, and focused overly on others, such as the possibility of PGS being a general tax-raising mechanism. The ability of the Treasury to raise the rate at some future point did not assuage the fears of the development industry that PGS would be a new form of general taxation, only part of which would be hypothecated to supporting development.

CIL or tariffs – little difference?

Ultimately, CIL was a face-saving exercise by government. While the moral argument case for the levy is strong, particularly in ensuring that development should contribute more – in 2005–6, only 14 per cent of development contributed through the Section 106 route (DCLG, 2010a: 8) – it is difficult to envisage what CIL can achieve that planning obligations could not. The shift from Section 106 as a mechanism to CIL would be more understandable if obligations were to be replaced entirely. However, the retention of the Section 106 mechanism for on-site issues results in a hybrid solution, and the HBF and BPF both argued that the approach should be either CIL or 106 but not both (*Planning*, 18 September 2009). The fear of the property industry was that, despite government assurances, local authorities would see CIL as being in addition to obligations.

The confusion at the heart of PGS reflected the ambiguous nature of the wider objectives for planning and was typical of other fuzzy, 'win–win–win' concepts deployed by New Labour. PGS, like 'sustainable development' and 'urban renaissance', promised an approach that would support growth through providing necessary infrastructure, overcome complexity and delay in the current arrangements, and provide a source of income for government based upon capturing a proportion of the uplift in land values resulting from public decisions. In reality, it failed to meet all three of these objectives, and the result is a system that does not fully address what are widely perceived to be problems with the system of obligations.

8
CONCLUSIONS

Introduction

The New Labour era officially came to an end on 11 May 2010, when Gordon Brown announced his resignation and the Queen asked David Cameron to form a government. The coalition between the Conservatives and Liberal Democrats was announced on 12 May, and on the same day Eric Pickles was appointed secretary of state for communities and local government. Within days the new government was beginning to unpick many of the policies and approaches of Labour and roll out its own agenda. Few had a clear idea what that agenda actually was and what it would mean for planning. Among immediate changes was the 'abolition' of Regional Spatial Strategies and density targets. In the longer term the coalition government intended to move towards a more local approach, incentivising communities to accept development rather than imposing it. How this would work was unclear, though a Localism Bill intended for late 2010 would provide some detail. The Conservatives' Green Paper *Open Source Planning* (Conservative Party, 2010) contained a range of ideas loosely linked to an overarching theme of 'localism'. The proposals seemed, on the face of it, to be positioned as the antithesis of Labour's approach. The analysis in *Open Source Planning* of the previous thirteen years was scathing and attacked the 'top-down', anti-democratic, antagonistic and ineffective nature of planning reforms under Labour. Such sentiments were echoed by some in the industry: 'We won't mourn the passing of density targets which, like most of the housing aspirations held by the last government, failed dismally to translate into any benefit for communities' (Liz Peace, BPF chief executive, 10 June 2010).

The change of government also allowed previous Labour ministers to reflect upon planning. John Healey, former communities and local government minister, acknowledged that Labour's approach had been too 'top down' (Healey, 2010). Wider reaction was divided, which was not helped by the vagueness of the proposals from the coalition. Some condemned the approach as the 'effective demolition of planning' (Hall, 2010: 268; Lock, 2010)' others saw it as an opportunity to achieve the original aspirations of notions such as spatial planning (Addison, 2010) or argue that the system need to be tinkered with rather than requiring wholesale reform (Lee, 2010). Others still saw chaos and uncertainty (Shepley, 2010). It is perhaps understandable that a change of government will give rise to some immediate reactions rather than more measured reflection on the past and future. However, in this concluding chapter I attempt to revisit some of the questions and issues raised towards the beginning of the book before offering a framework in order to understand better the nature of change in planning and then briefly looking at the direction of change since May 2010.

New Labour revisited

In Chapter 2 I set out an approach to understanding Labour and planning based on three questions: *What* New Labour?, *Which* New Labour? and *When* New Labour? How does this understanding fare, and is it possible to develop it in the light of the subsequent empirical chapters?

What New Labour?

The discussion in Chapter 2 highlighted the lack of agreement over the ideological origins and orientation of New Labour. Depending upon the particular focus or, more significantly, the position of those making the assessment, New Labour was a continuation of Thatcherism, a branch of Northern European social democracy, a *sui generis* blend of ideas for 'new times' or a form of pragmatism or opportunism that defied easy classification. Is it possible to identify coherence or discern a New Labour ideology from the experiences of planning covered in the previous chapters? The simple answer is no. There was an underlying belief that the system could be used to achieve objectives or themes such as social justice, environmental stewardship and economic competitiveness and that the processes and outcomes of planning could be shifted and 'improved'. But this does not amount as much to an ideology as to a 'wishful orientation'. There was no 'big picture' but a succession of vague multiple themes or discourses that were open to interpretation. This ambiguity provided the basis for local variability, disagreement and conflict which might have been avoided had there been a clear, guiding, overall direction. The lack of any steer to change also impacted upon the means and approaches employed, which were derived largely from expert panels or commissions. Peter Burnham has termed Labour's tendency to govern at a distance as 'depoliticisation' (2001). Ministers maintain the overall direction of policy while benefiting from the ability to shift responsibility for unpopular decisions. The downside is that the overall direction or objectives remain confused, particularly when such depoliticisation derives from different ministers and even from different government departments. The Rogers, Barker and Killian–Pretty reports, to name a few, underpinned changes to planning at different times, though each had a different view of the purpose of planning and how it should be approached. This was a form of 'managerial pragmatism' (Finlayson, 2009) and amounted, at times, to a 'hotch potch' of initiatives and programmes (Lowndes, 1999; Stoker, 2000). There is evidence to support the notion that Labour's approach to planning was eclectic and incremental: How could the pursuit of a policy that was almost universally felt to reduce development (the Planning Gain Supplement – see Chapter 7) be reconciled with the clear intention of increasing development and, in particular, housing?

In part, such contradictions arose as much through a lack of understanding of the nature of planning as from a lack of overall direction. Labour's obsession with 'positive messages' and its zeal to eradicate the negative, particularly in the early years, precluded and avoided conflict. Contradictions also arose from the need to depend upon others to effect change, particularly at the local level. The lack of understanding of planning occurred at a number of levels, though the assumption that strengthening the statutory link between development plans and development control would unproblematically increase speed and certainty in the process ran against the experiences since 1991, when the plan-led approach was introduced. However, it was the scope of discretion and autonomy found within the planning system and processes beyond Westminster that filled the policy vagueness and change and helped shape '*What* New Labour?' While there was a superficial confluence of economic, social and environmental concerns within abstract, national policy aspirations, the interpretation necessary for implementation led to disparate approaches that reflected local priorities and objectives.

Which New Labour?

The themes running through the approach to planning during the New Labour era exhibited schizophrenia between, on the one hand, the desire to create inclusive, socially, environmentally and economically integrated communities based around multi-scalar partnerships and processes of governance and, on the other, a clear commitment to growth, development and economic competitiveness through, for example, top-down housing targets and an emphasis upon speed and certainty. The lack of resolution of the purpose of planning was clear from the experiences in urban regeneration and spatial planning, as well as in the ambiguity over whether the Planning Gain Supplement was a tax-raising mechanism. Such ambiguity echoes the different strands of thought within the party and government over the role of the state in a global market economy. These tensions were not unique to Labour and, as discussed in Chapter 1, the New Right approach was also bifurcated between the deregulatory, liberal strand and the centralising, authoritarian tenet. Nevertheless, addressing '*Which* New Labour?' from the competing understandings within government was not helped by the incremental and reactionary addition of policies and objectives that were driven more by events than ideology. The role of planning, for example, in helping tackle terrorism and addressing childhood obesity (see Chapter 3) were part of the significant expansion of objectives noted by the second Barker report (Barker, 2006b) and derived from reactions to events rather than an understanding of what planning could deliver.

The related element in understanding the nature of New Labour from the experiences of planning concerns the interface between the aspirational objectives around consensus-based spatial governance and the use of a system designed for different times and focused upon conflict management rather than consensus-building. The lack of a fundamental rethink around the mechanisms of planning exhibited, at best, a misunderstanding of how change could be effected through planning and betrayed, at worst, the absence of a steering ideology to guide its approach. The introduction of the new system of development planning, discussed in Chapter 5, was initially intended to speed up and simplify matters and provide greater certainty. Ironically, there was a clear foundation and logic to the proposals, which were linked to a market-supportive, deregulatory vision of planning. However, the shift in purpose of the new system from 2002 did not include a shift in the approach and mechanisms to development planning – Regional Spatial Strategies and Local Development Frameworks were seen as suitable tools for a vision of planning that was more market driven and an approach that was multi-tiered and concerned with sustainable growth and objectives.

When New Labour?

Notwithstanding the problems with identifying boundaries and periods against inevitable fuzziness, Chapter 2 highlighted distinctive eras to planning under Labour. The broad argument was that, initially, Labour had no clear view on planning that was distinguishable from that of the previous Conservative government. This changed from around 2000, when a more market-led approach was in the ascendency. This was eclipsed by more socially and environmentally aware concerns until the onset of the credit crunch and recession from 2007, when a greater concern with delivery and economic competitiveness emerged.

The evidence from the four empirical chapters does not significantly challenge this overall view, though it does temper it. If we take the approaches to development control and development planning we can see that they were regarded and tackled very differently under Labour. The Planning Delivery Grant (PDG) was introduced in 2003/4. As an approach it was underpinned by a view of development control as a largely technical process that could be better 'managed' (i.e. made quicker and more certain) if local authorities had the will. This 'managerial pragmatism' contrasted with the parallel approach to development planning, also introduced from around 2003/4, that, first, equated 'planning' with development planning and, second, gave this element of planning a set of objectives that echoed those of New Labour as a whole. Planning – or, more accurately, development planning – was charged with delivering a progressive agenda built upon growth, social inclusion and environmental stewardship. Both themes existed in parallel and were underscored by very different approaches to and understandings of the objectives and mechanisms of planning. This does not undermine the idea of distinct eras but does temper the notion by emphasising the need to distinguish between dominant and recessive governing ideologies within an era. For example, while the period 2004–7 was characterised by an emphasis upon inclusion and sustainability, there was a less obvious theme of economic competitiveness and growth that influenced some areas and approaches to planning. During the period 2007–10, as the concern with delivery and the impacts of the recession began to overshadow other themes, this recessive concern began to be more influential and dominate.

Explaining New Labour's approach to planning

The experiences of planning do not provide much clarity on the ideology of New Labour. The rather unsatisfactory outcome from the above analysis does, however, raise a number of questions around the motors of policy and change. Part of the problem in looking for a coherence or ideology is that such concepts might themselves be dated and no longer applicable (Powell, 2000). Another issue is that we might be looking in the wrong place. One theme from the New Labour era common to the approaches taken concerns 'regulation' (the centrality of the role of the state), as opposed to 'deregulation' (the centrality of the market). Spatial planning, urban regeneration, targets and infrastructure provision in relation to growth and development were organised and arranged around the creation of new processes, institutional arrangements and funding mechanisms across the public and private sectors. To some this might imply a reversal of the deregulatory instincts of the New Right. Yet the underlying purpose and objectives of a 'state-led approach' should not necessarily be taken at face value, particularly as a major concern of Labour was clearly with economic growth and competitiveness. One way in which New Labour's approach to planning would be better conceived, and which helps us understand the fusion of state and market, is as a form of neoliberal spatial governance.

There is no shortage of New Labour analyses that link it to a form of neoliberalism (see, for example, Hay, 1999, and Callinicos, 2001). Since the collapse of the communist regimes in the Soviet Union and Eastern Europe, the dominant government ideology has been liberalism and the primacy of free markets and individual freedom. This economic configuration and its globalisation requires that the state, if it is to compete and thrive, should 'create and preserve an institutional framework appropriate to such practices' (Harvey 2005: 2). There is not one neoliberalism (Gamble, 2009), though within the UK it has included a range of manifestations, among them the privatisation or commodification of the state and, in particular, the local state (Stoker, 2004), a rescaling of state activity to support economic activity (Jessop, 2002b; Brenner and Theodore, 2002) and a dominant emphasis within public policy upon economic growth and competitiveness (Raco, 2005).

How such practices evolve at the state level will vary (Peck and Theodore,

2007; Brenner and Theodore, 2002). Varying neoliberal solutions focuses attention on the ways in which the state, at all levels, creates and destroys modes of governance in a constant and impatient search for new, more efficient and effective mechanisms that overcome the contradictions and crises of liberalism and globalisation. For our purposes it is possible to identify *temporal* variegation as well as *spatial and scalar* variegation in neoliberal governance. Temporal variegation highlights the evolution and variation in the search for more effective strategies of neoliberal governance. Here, the broad differences and evolution of neoliberalism can be identified in the change from crude, deregulatory strategies in the 1980s to a more market-supportive re-regulation during the 1990s which privileged public–private partnerships and targeted public intervention to help address the contradictions and outcomes of previous periods of economic and social change (Fuller and Geddes, 2008: 256). Spatial and scalar variegation emphasise how neoliberal strategies vary through space and across scales and amount to what Brenner (2004) terms 'state spatial projects', which seek to bring internal coherence and functional coordination. According to this view, the rescaling of planning and other state activities in recent years is an attempt at a better coordination of state activities among diverse locations.

How does the adoption of a variegated neoliberal understanding help explain the experiences of planning during the New Labour era? One of the issues that any perspective needs to address is the existence and influence of multiple discourses and policies around planning that, on the face of it, seem to be at odds with the idea of a dominant, market-led approach. As Tickell and Peck (2006) point out, neoliberalism will often unfold in a pragmatic and variable way on account of the interaction between neoliberal principles, which can translate into a range of policy and governance approaches, and the existing institutional arrangements and social, economic and political landscapes (see Brenner and Theodore, 2002). As a consequence we should not necessarily assume that the particular strategy of the state will be consistent with broader, neoliberal tenets, and hybrids between neoliberalism and other philosophies will inevitably exist (Raco, 2005). This is compounded in the UK by the discretion afforded to local planning authorities and the emergence of different planning styles, as discussed in Chapter 3. Nevertheless, New Labour's approach to planning accelerated the evolution of and changes between discourses as well as moving away from one dominant discourse to a range of competing discourses.

Neoliberalism helps explain the underlying approach of New Labour and its boundaries as well as the inherent tensions and contradictions of the different streams. Under this perspective planning and spatial planning can be seen as a form of neoliberal spatial governance providing the necessary strategies and institutional fixes in order to legitimise and facilitate growth. What arguably marks out New Labour's approach is the extent to which growth was seen as necessary to deliver social democratic objectives around social inclusion and social justice, as well as the mechanisms through which this strategy was deployed, particularly the use of all-embracing, positive discourses to help generate a consensus around development and eradicate conflict and opposition. Understanding planning as a form of neoliberal spatial governance helps explain some of the approaches taken under New Labour, but it also stresses the importance of spatial and sectoral variegation. How such ideas are mediated and implemented by actors and networks cannot be 'read off' from the motors of change operating at other, higher scales. Such implementation will itself vary depending upon the scope of change, as while there were some areas where Labour did have a clear intention to effect change, there were others that were left very much to the local level.

In Chapter 3 I discussed how New Institutionalism provided a cogent framework for understanding how change was interpreted, mediated and implemented. New Labour understood governance from an institutionalist perspective, emphasising the networked, non-hierarchical flows between a range of bodies and actors that underpinned its issue-based approach to policy and the focus upon vehicles of coordination and integration such as Local Strategic Partnerships, Community Strategies and Local and Multi-Area Agreements. As Lowndes rightly notes, defining the conceptual framework for institutional emergence is only the first step: 'any test of the robustness of the concepts must be an empirical one' (2005: 306). Lowndes's emphasis upon 'rule sets' highlights the 'stickiness' of established practices and customs (political, managerial, professional and constitutional) and the dynamic and evolving relations and networks between them. If her framework helps explain inertia within local governance, arrangements then Schmidt's approach (2008) privileges and helps explain change. In her view, various discourses provide institutionalised structures of meaning that channel political thought and action. Discourses help identify issues and shape and galvanise responses across policy sectors, networks and groups (e.g. advocacy coalitions, epistemic communities, etc.) and scales. Crucially, to be persuasive, discourses need to be consistent and coherent – though they can also be vague, which is often useful in developing appeals.

Both stability and change frameworks can be applied to Labour's approach to planning and spatial governance. Clearly, the complexity and evolution of planning over a thirteen-year period and the intersection of differing streams of policy through time and across sectors mean that it is possible to point to examples of change *and* stability. Thus, both frameworks of understanding help identify and understand various experiences better. An institutionalist perspective helps 'makes sense' of New Labour's approach to planning, though, in turn, it is also necessary to reflect upon the institutionalism itself.

- The various New Labour planning discourses were vague *and* positive, encouraging 'buy in' to various growth-centred strategies. Two consequences stand out. First, it was not until detailed development schemes emerged and the various interpretations of discourses such as sustainable development 'became real' that difference and conflict arose. Second, there were implications for the ways in which 'institutional entrepreneurs', as Lowndes (2005) terms them, operated. The vagueness and contradictions of the discourses provided institutional entrepreneurs with the opportunity to draw upon what Schmidt (2008: 311) terms 'background' discourses and influences, which include professional codes, doctrines, local cultures and preferences. The scope for discretion in planning, discussed in Chapter 3, is considerable and enables a variable geometry of practice and policy to emerge. This had positive and negative consequences, allowing for both change and resistance to change. On the positive side, the 'soft spaces' and 'fuzzy boundaries' around the Cambridge sub-region sought to work around the problems of planning jurisdictions that were not functional for the purposes of growth. On the negative side, the extension of performance indicators and management in development control provided incentives and encouragement to focus on the process rather than the outcome, leading to new rules (e.g. refusal of applications rather than negotiation, etc.).

- Schmidt (2008) advances the notion that her three levels of ideas – policies, paradigms and philosophies – come in two types: cognitive and normative. Cognitive ideas relate to 'what is and what to do', while normative ideas include 'what is good or bad' about what we should do. The emergence and use of spatial planning highlights how this distinction does not capture the complexity of how this particular notion structured both the cognitive and the normative. The post-political analysis outlined in Chapter 3 and the experiences of approaches such as spatial planning (Chapter 5) would suggest a conflation of the two. Spatial planning was a persuasive discourse aimed at 'winning the argument' around growth and involved both cognitive (process- and outcome-related elements – 'what to do and how to do it') and normative elements (why spatial planning was 'superior' to regulatory planning). A critical part of this 'winning the argument' involved a sophisticated but misleading back-story around the origins and wider lineage of spatial planning, though the most significant element was the way in which spatial planning was promoted not only by government but also by the wider planning community, both practitioners and academics.
- Ostrom's (1999) distinction between 'rules in use' and 'rules in form' echoes some of the experiences of change in planning (Lowndes, 2005: 292). The strategic behaviour around and the perverse outcomes of the Planning Delivery Grant highlight how superficial elements of the 'modernisation' of planning really were. The experiences of planning during the Labour era point to a more nuanced and complex understanding of formal and informal rules. The role of discretion and the influence of different actors and multiple and conflicting objectives mean that formal rules, in Lowndes's sense, are not black letter law but are themselves open to interpretation and creativity. The indefinite nature of government objectives for planning expressed through national policy statements and supporting information was such that plans and strategies were themselves less than definite. However, this is not unique to the Labour era: plans and strategies in the UK tradition have traditionally been indicative rather than prescriptive. What it does highlight is a continuum or

gradient of informal rules rather than a more binary formal–informal distinction.
- During the earlier period of New Labour the approach to planning largely followed the trajectory of the previous Major government. Policy and institutional evolution was 'normal'. From 2002 a more punctuated evolution commenced, accelerated by a range of demands upon and objectives for planning combined with the introduction of a new approach to development plans and a development boom. Following Torfing (2001), we could point to the elasticity of existing rule sets, which can account for and cope with change through reinterpretation of practices and policies. However, this is possible only up to a point. Institutional dislocation occurs when the limits of elasticity are breached (Davies, 2004). At this point there is an opportunity for new rule sets, advanced by institutional entrepreneurs, to emerge. The punctuated evolution of planning had a number of outcomes. The emergence of a myriad of 'soft spaces' combined with fuzzy boundaries, as planning and planners sought to roll out spatial planning across scales and sectors and administrative boundaries, may have been foreseen, though it had a range of unforeseen consequences that worked against other objectives: the need for a complex assemblage of bodies and committees to coordinate and deliver growth blurred accountability and transparency. This observation on the wider impacts of partnership and networked governance is beginning to gain traction in other fields (see, for example, Geddes, 2006). Sorenson and Torfing (2004: 6) challenge the contention that 'governance networks are to an increasing extent seen as an effective and legitimate form of societal governance'. They suggest that network governance arrangements might, on the one hand, result in a positive set of outcomes, including:

- enhancing a vertical balance of power within the governance system, establishing a link between top-down representative democracy and bottom-up participative democracy;
- serving as a medium for enhancing political empowerment and trust;
- improving governance efficiency and outcome legitimacy.

On the other hand, there are potential dangers inherent in network governance, which may:

- undermine political competition and autonomy, reducing discursive contestation of political and policy alternatives;
- make political and policy processes less transparent and public;
- undermine both elected politicians and community participation.

From the above engagement between the experiences of planning under Labour and institutionalist analysis we can begin to develop some broader understandings around the nature of planning and change. The main point concerns the emergence of new significant rule sets that are meshed onto existing and evolving sets. The fusion and resolution of such sets and wider institutional emergence will vary depending upon the locality and the issue. The result is a range of tensions between new and existing rule sets which includes but are not limited to the following.

- *Hierarchical versus spatial*   Despite the notion of networked governance, there is still a strong hierarchical element of policy cascade in planning from national to local level. This approach is meant to ensure consistency while allowing for flexibility for different areas to develop their distinctive policies. The more spatial rule set emphasises coordination and integration across and between scales. The problem of reconciliation was not helped by conflicting messages from government between 'top-down' targets and performance and the more locally led spatial emphasis on coordination and integration.
- *Political versus bureaucratic*   Planning employs a range of processes, from political/distributional (e.g. how much development and where) to quasi-judicial (e.g. making decisions that affect rights) and bureaucratic (e.g. making decisions quickly, consistently and fairly). Spatial planning was underpinned by a largely bureaucratic view (i.e. that integration and coordination between sectors and across scales was a largely technical/managerial process) and ignored the political, distributional element at the local and sub-regional level. Similarly, the plan and permission nexus was perceived as a bureaucratic rather than a political link.
- *Sectoral coordination versus professional/legal silos*   Sectoral integration and coordination as an objective was backed up with mechanisms and resources through, for example, Local and Multi-Area Agreements. However, sectors such as housing, transport or economic development also have specific, regulatory and professional foundations that require separate processes and draw upon distinct professional doctrines. While integration at a strategic policy level was being encouraged and pursued, the regulatory and professional dimension continued to provide 'institutional stickiness'.
- *Service delivery versus 'place-making'*   Performance targets in development control and, increasingly, development planning emphasised the need to increase certainty and make quick decisions and strategies while, at the same time, 'place-making' and spatial planning were recognised as open-ended, quality-based processes that did not require or even involve concrete outputs such as a plan or strategy. How these two rules sets were brought together, both intellectually and practically, was left largely unresolved.

The emergence of competing rule sets also raises the likelihood that institutional change might be superficial – for example, 'playing the game' of change and making decisions and strategies based upon other rule sets. Where new rule sets emerge and discretion is high we can expect 'institutional entrepreneurship' to come into play and either advance new sets or focus upon other, more established sets. In such contexts, 'institutional remembering' can be used to give a new rule set a convenient and useful back-story from more established sets (e.g. spatial planning and as a return to older, established forms of planning), 'institutional borrowing' can draw upon ideas or discourses from other rule sets (e.g. the application of private-sector management discourses and processes) or 'institutional sharing' can take the form of sharing experiences and examples between sectors and across jurisdictions (e.g. the establishment of the web-based good practice, such as the Local Government Improvement and Development website and network).

Planning as neoliberal spatial governance

Bringing together the two elements – understanding New Labour's approach to planning as an evolving form of neoliberal spatial governance, one the one hand, and an institutionalist analysis of change, on the other – provides a framework within which to understand the experiences, successes and failures of planning through time. Table 8.1 sets out a heuristic process that provides one way of conceiving of this relationship. Within a neoliberal philosophy there are a range of paradigmatic discourses that seek to frame debates, policies and action. Linked to these discourses are a range of policies that reflect the ambiguity of the paradigms and the wide scope for interpretation. Thus some policies are closely aligned with the paradigm while others are less so. Paradigms evolve with regularity, echoing what Massey (2005) has termed 'temporary permanences' within public policy under New Labour. In planning, the shift in paradigms, which is less abrupt than represented in Table 8.1, arose from competing ideologies gaining transitory dominance. Thus, the view of planning as a supply-side constraint upon growth and the economy was the dominant paradigm from 2000 to 2004, though this was supplanted by a paradigm based upon planning as a form of 'governance glue' that coordinated and integrated across and between scales and sectors. Dominant paradigms do not eradicate other competing paradigms but have temporarily either supplanted or incorporated them. Thus the sustainable development paradigm sought to incorporate competing social, environmental and economic discourses through synthesis and the persuasive idea that spatial planning could deliver 'win–win–win' outcomes. This paradigm broke down and was itself replaced for the reasons discussed in Chapter 5. What is worth noting, however, is that the ascendency of paradigms occurs, as Schmidt (2008) argues, when one discourse becomes more persuasive in addressing the problems and issues at hand. However, as the experience of New Labour and planning highlights, paradigms and policies do not change instantaneously or seamlessly. This is because there is an element of choice or discretion involved and because of lags in the adoption and implementation of policies. One outcome is that there are a range of policies that can meet the requirements of the dominant paradigm, some of which will contradict or oppose the others. Another is that, in practice, the policies row in Table 8.1 should be moved to the right to point out the lag and mismatch with the dominant paradigm.

TABLE 8.1 A New Institutionalist understanding for change in planning

Philosophy	Neoliberalism						
Dominant paradigms	**Project-led** Deregulation and centralisation.	**Plan-led system** Acceptance of the need for strong policy framework for market to work.	**Plan-led system and devolution**	**Urban renaissance and economic growth**	**Sustainable Communities** (encompassing urban renaissance, economic competitiveness and environmental concern with climate change). **Spatial planning** as process and output Speed and certainty in development control.	**Growth and delivery** (development management introduced to address the problems of spatial planning paradigm and reforms to the post-2004 system of development plans).	**Localism and Open Source Planning** Greater responsibility for localities to shape their places within pro-market context.
Policies	Deregulation and experimental spaces, e.g. UDCs and SPZs. Centralisation and reduction of discretion.	Consistency and speed. Greater enforcement of national priorities through regional government offices and introduction of plan-led approach in 1991.	Competiveness and growth. Planning policy overshadowed by new Regional Development Agencies. Devolution enables policy 'experimentation'.	Local Development Frameworks, Regional Spatial Strategies as tools of deregulation/ increased certainty. Plans subsumed within Community Strategy prepared by Local Strategic Partnership. Business Planning Zones proposed. Brownfield emphasis and range of mechanisms and strategies to achieve 'sustainable development', e.g. density targets.	Integration and coordination at functional planning scales (e.g. Local Area Agreements). Local Development Frameworks, Regional Spatial Strategies as tools of sustainable communities and spatial planning. Planning Delivery Grant and performance targets. Barker reviews. Eco-towns. Urban Development Corporations/delivery vehicles.	Community Infrastructure Levy. Infrastructure Planning Commission. Killian-Pretty review. Local Democracy, Economic Development and Construction Act 2009. Reorientation of PGs to delivery. Review of Regional Spatial Strategy housing targets.	Abolition of Regional Spatial Strategies. Greater use of financial incentives to promote acceptance of growth at the local level. Local Development Orders to deliver simplified regime. Presumption in favour of development if plan in place.
Scales	National dominant through diminution of local discretion.	National/regional Illusion of local control.	Increased emphasis upon regions and nation regions within loose national policy framework.	RDAs increasingly influence spatial strategies and communities grapple with the evidence base for plans.	Emergence of soft spaces and fuzzy boundaries as localities explore functional spaces to achieve integration and coordination.	Stronger central direction in directing policy and delivery.	Local.
Era	1979–91	1991–97	1997–2000	2000–04	2004–07	2007–10	2010–?

In other words, by the time the policy implications of a particular dominant paradigm emerge, the paradigm may itself have evolved and changed.

The experience from the Labour years demonstrates how this mismatch occurred and how it effected the nature of change to planning. Spatial planning through 'soft spaces', involving a range of bespoke, partnership-based delivery bodies, was characteristic of and necessary to the post-2004 approach. However, such bodies and approaches took time to establish and gain legitimacy and traction. Cambridgeshire Horizons, discussed in Chapter 5, was established in 2004 as, *inter alia*, a sustainable communities delivery vehicle, though at the point that the growth plans were to be realised it was undermined by the property recession and public-sector funding cuts. The paradigm within which Horizons existed shifted from the Sustainable Communities growth agenda to economic growth and delivery to the current deregulatory and localism paradigm (Table 8.1). The latter paradigm would allow partners within Horizons to withdraw their cooperation as the ethos shifted from cooperative, spatial governance to more individualistic regulation. Horizons has had to adapt to the changing paradigms and policies and, while such flexibility is necessary and welcome, it does have implications for effecting change, which takes longer to achieve than the paradigm within which it exists.

The post-Labour era

If planning as a form of neoliberal spatial governance demonstrates anything it is that paradigms and policies evolve, highlighting the adaptability of planning to accommodate a range of paradigms and policies. Nowhere is this better demonstrated than in the shift towards the end of the New Labour era and the evolution from 2010. There was a *fin de siècle* feel from 2008 onwards, as Labour's approach to planning oscillated from incessant reform to resignation that the coming general election was lost. The publication of the Conservatives' *Open Source Planning* (Conservative Party, 2010), following the Bow Group's earlier *Our Towns, Our Cities: The Next Steps for Planning Reform* (Cuff and Smith, 2009), was met largely with a mixture of disappointment, apathy and inevitability from a sector that had experienced reform upon reform over the previous decade, often without a clear understanding of what was to be achieved, how the system worked, and how change was to be effected in a complex and conflict-ridden environment.

If few people knew what *Open Source Planning* would mean in practice, one thing that could be agreed upon was that, to paraphrase Tony Crosland thirty-five years earlier, the party was over. The combination of the need to reduce the national debt combined with an ideologically hostile opposition that could point to the resource increases in planning without significant progress or performance improvement indicated there would be a 'roll back' of the state dressed up in the popular language of localism. It mattered little that what *Open Source Planning* proposed was actually inchoate and incoherent.

Peter Hall once called for an 'essay in non-plan' as a way of igniting and facilitating the growth and spirit of Hong Kong in Britain's inner cities (Banham *et al.*, 1969). This directly influenced the Conservative Party when in opposition in the 1970s to propose Enterprise Zones and, eventually, Simplified Planning Zones (Allmendinger and Thomas, 1998). On 22 June 2010, in the emergency budget, the coalition government announced the establishment of a 'simplified planning consents process' in certain areas. Along with the abolition of Regional Spatial Strategies and the introduction of presumption in favour of sustainable development that conforms with national and local policies, the thrust of the proposals was 'less is more', but with a greater say in the 'less' going to local communities in certain circumstances. 'Localism' was a dominant theme in *Open Source Planning*, though it was tightly circumscribed. Local communities needed to be proactive in resisting development: 'unless they [local planning authorities] use their local plans explicitly to rule out particular types of development in specific areas, the planning system will automatically allow applications to be approved' (Conservative Party, 2010: 13). This approach would encourage local authorities to prepare plans – but plans that were regulatory rather than spatial.

Many organisations also published their own manifestos for planning in the run up to the 2010 election, and some organisations did not waste any time in developing future approaches to planning which, to varying degrees, criticised the erstwhile Labour regime. The British Property Federation (BPF) (2010) argued that the system had become bureaucratic and peripheral to decision-making, while the Town and Country Planning Association (TCPA) (2010a) argued for the need for regional or strategic planning. The TCPA challenged some of the analysis, assumption and policies in *Open Source Planning* but sought to work with the spirit of the coalition's concerns through influencing the emerging agenda (TCPA, 2010b). The more mainstream business view was typified by the British Chambers of Commerce (2009), which criticised the slowness of the system and the uncertainty to applicants. The Planning Officer's Society (2010) and the Royal Town Planning Institute (RTPI, 2010) also issued their own planning-specific manifestos. The Planning Officers focused on the strategic planning vacuum that would result from the abolition of Regional Spatial Strategies, while the RTPI argued that the main elements of the system were largely in place and reform was unnecessary. It further maintained that, although some changes could be introduced to ensure that communities engaged better with the process, the system needed more resources and a higher profile within local authorities to ensure that it could achieve what the government sought. Many of the arguments from the pro-business lobby could have been written thirty years earlier: the British Chambers of Commerce manifesto, for example, used the time taken to secure planning permission for Terminal 5 at Heathrow as part of its argument that planning was bureaucratic, even though the decision had been issued nearly a decade earlier and the Infrastructure Planning Commission had been established to help address the problem. Yet the overall view of these analyses was that Labour's reforms, however well intentioned, had misunderstood the nature of the development process and the intricacies of the planning system and had failed to have a coherent and consistent approach.

On balance, Labour's reforms had mixed success in effecting change. Part of the impact was on account of poor, reactionary and ill-thought-through ideas (e.g. the Planning Gain Supplement), but there were approaches that were generally welcomed (e.g. the urban renaissance agenda and spatial planning) but were either undermined by how they were approached (e.g. density targets) or were undercut because other aspects of planning had not been addressed (e.g. spatial planning and development control). Overall, the sheer extent, vagueness and rapidity of change and initiatives were overwhelming.

It is perhaps understandable that the 'less is more' narrative and approach post-New Labour is superficially attractive. However, the flipside is uncertainty, confusion and inaction across the planning and development sector. As the chairman of the Home Builders Federation, Stewart Basely, said:

> We urgently need clarity on housing planning policy if the government is to deliver its pre-election pledge to build more of the homes it recognises we need. We all expected the regional plans to be scrapped, but we now need direction on how we are to move forward. We have an acute housing crisis in this country, approaching a shortfall of a million new homes. We just cannot afford a period of confusion to reduce house building still further at a time when we are already building at the lowest level for many decades.
>
> *(HBF press release, 20 May 2010)*

At the time of writing (summer 2010) the vacuum of policies and the uncertainty over the future of planning was beginning to dawn upon the sector. A promised Decentralism and Localism Bill to enable many of the proposals would be unlikely to reach the statute books before 2012. In the meantime, the 'death by a thousand cuts', through both significantly scaled-back public-sector spending and the cumulative impact of many policy announcements and intentions, would leave little upon which to build – which is perhaps the real intention.

A number of points arise from analysing the changes from the perspective of Table 8.1 and planning as neoliberal spatial governance. First, the coalition's approach is not a return to the crude deregulation of the 1980s. In an era of climate change, and against a backdrop of a more sophisticated understanding of and approach to economic globalisation than existed thirty years ago, the market-led framework is wrapped in a positive discourse of localism. While localism is as vague as 'sustainable communities' and 'urban renaissance', it provides the mainly conservative and Conservative local planning authorities with an incentive to produce plans, if only to manage (i.e. stop) development. In the immediate aftermath of the election, a number of authorities that had been resisting housing growth through legal challenges to Regional Spatial Strategies withdrew their LDF Core Strategies. In other words, *Open Source Planning* is not an essay in non-plan but an essay in restrictive plan. Second, clearly the nature of the plan will change. In the recent past, detailed plans have been positive in identifying where and in what circumstances development will occur. There has also been a core of policies, either criteria based or spatial, that say where development should not occur. If the presumption in favour of sustainable development is introduced as proposed, then it will require a plan. Development will be permitted if it is in accordance with the plan. This takes the plan-led approach one step further, towards a more prescriptive, zoning-based system. It will change the nature of the plan itself. There will be a tendency to have 'minimal' plans that allow little in order to maintain greater local control. Alternatively, as the experience of simplified planning zones highlighted, combined plan and permission approaches were circumscribed by large numbers of conditions and caveats in order to restrict and manage all possible eventualities. Plans are likely to become risk averse and take over some of the bureaucratic and legalistic character of the development control process they will impinge upon. Finally, the use of financial incentives to promote growth and development in some areas has long been touted, and experiences in other parts of Europe demonstrate that they could have some impact. However, the fiscal and governance arrangements in the UK differ significantly, meaning this is less likely to make a difference. However, what is perhaps not as clearly understood is that under such an approach local communities could in effect pay *not* to accept development. If the council tax in an area would be reduced by, say, £100 per annum per household by accepting development and the financial incentives that come with it, then a community could decide that not having development would be worth that much to them. Those more affluent areas, particularly though not exclusively in the south of England, that have resisted development may be prepared to pay to do so in the future. In some ways such an approach is a challenge to the neoliberal, growth-led philosophy, and it is notable that developers and investors have largely welcomed the incremental changes proposed by the coalition government but have expressed concern over the more fundamental proposals.

These are not the only issues to arise and, given the uncertainty and the time that the changes are likely to take, there will inevitably be others. However, there are some parallels between the coalition and New Labour. Planning figured very little in the first three years of the Labour government and, similarly, seems not to be a priority in the coalition. In Labour's case, the vacuum was filled by a variety of practices, though, in retrospect, resulted in a missed opportunity, particularly given the time taken to effect change. A related lesson that the coalition could learn concerns how long it took to implement reform of development planning. Six years after the 2004 Act the majority of local planning authorities still had not adopted their Local Development Framework Core Strategies. Such experiences would seem to be overlooked. Similarly, the lessons from the 1980s deregulated form of planning appear to be lost. Every government evinces a degree of hubris in considering that its approach has better understood the issues and the policies needed to address them. Although cloaked in the language of diversity, localism and information technology (open source software allows users to change and improve the code to meet their own needs), the coalition's approach is a shift away from 'rolled-out' to 'rolled-back' neoliberalism. In a favourite mantra of Tony Blair when he was persuading the Labour Party to accept the reforms to Clause 4 of its constitution, the coalition government may be living the past and not learning from it.

NOTES

5 Spatial planning

1 The work on Thames Gateway was undertaken with Graham Haughton, Dave Counsell and Geoff Vigar.
2 'Integrated spatial planning, multi-level governance and state rescaling' (ESRC grant 000230756), with G. F. Haughton, D. Counsell and G. Vigar.

6 Hitting the target and missing the point

1 www.planningportal.gov.uk/england/professionals/en/1115315772911.html.

REFERENCES

Addison, L. (2010) Why we need to make spatial planning work, *Town and Country Planning*, 79(5).
Adler, M. and Asquith, S. (1993) Discretion and power, in M. Hill (ed.), *The Policy Process: A Reader*. Brighton: Harvester Wheatsheaf.
Albrechts, L. (2006a) Shifts in strategic spatial planning? Some evidence from Europe and Australia, *Environment and Planning A*, 38(6): 1149–70.
Albrechts, L. (2006b) Bridge the gap: from spatial planning to strategic projects, *European Planning Studies*, 14(10): 1487–500.
Alexander, D. (2007) A resurfacing tension, *Town and Country Planning*, June/July: 213–17.
Alexander, E. R. and Faludi, A. (1996) Planning doctrine: its uses and implications, *Planning Theory*, 16: 11–61.
Allen, J. and Cochrane, A. (2007) Beyond the territorial fix: regional assemblages, politics and power, *Regional Studies*, 41(9): 1161–75.
Allmendinger, P. (1998) Simplified Planning Zones, in P. Allmendinger and H. Thomas, *Urban Planning and the British New Right*. London: Routledge.
Allmendinger, P. (2006) Zoning by stealth? The diminution of discretionary planning, *International Planning Studies*, 11(2): 137–43.
Allmendinger, P. (2009) Explaining the paradox of performance improvements and delay in development control, *Urban Design and Planning*, 162(2): 79–86.
Allmendinger, P. (2010) Transaction Costs, Planning and Housing Supply. London: RICS.
Allmendinger, P. and Ball, M. (2006) *Rethinking the Planning Regulation of Land and Property Markets: Final Report*. London: ODPM.
Allmendinger, P. and Haughton, G. (2007) The fluid scales and scope of spatial planning in the UK, *Environment and Planning A*, 39(6): 1478–96.
Allmendinger, P. and Haughton, G. (2009a) Soft spaces, fuzzy boundaries and metagovernance: the new spatial planning in the Thames Gateway, *Environment and Planning A*, 41: 617–33.
Allmendinger, P. and Haughton, G. (2009b) Critical reflections on spatial planning, *Environment and Planning A*, 41: 2544–9.

References

Allmendinger, P. and Tewdwr-Jones, M. (1997) Post-Thatcherite urban planning and politics: a major change? *International Journal of Urban and Regional Research* 21(1): 100–16.

Allmendinger, P. and Tewdwr-Jones, M. (2000) New Labour, new planning? The trajectory of planning in post-New Right Britain, *Urban Studies*, 37(8): 1379–402.

Allmendinger, P. and Tewdwr-Jones, M. (2009) embracing change and difference in planning reform: New Labour's role for planning in complex times, *Planning Practice and Research*, 24(1): 71–81.

Allmendinger, P. and Thomas, H. (1998) *Urban Planning and the British New Right*. London: Routledge.

Ambrose, P. (1986) *Whatever Happened to Planning?* London: Methuen.

Amin, A. and Thrift, N. (1995) Institutional issues for the European regions: from market and plans to socioeconomics and powers of association, *Economy and Society*, 24: 41–65.

Amin, A., Massey, D. and Thrift, N. (2000) *Cities for the Many, Not the Few*. Bristol: Policy Press.

Argent (2001) *Principles for a Human City*. London: Argent.

Atkinson, R. (2004) The evidence on the impact of gentrification: new lessons for the urban renaissance? *European Journal of Housing Policy*, 4(1): 107–31.

Audit Commission (1992) *Building in Quality: A Study of Development Control*, Local Government Report no. 7. London: Audit Commission.

Audit Commission (1994) *Watching their Figures: A Guide to the Citizen's Charter Indicators*. London: Audit Commission.

Audit Commission (2002) *Development Control and Planning*. London: Audit Commission.

Audit Commission (2006a) *Securing Community Benefits through the Planning Process: Improving Performance of Section 106 Agreements*. London: Audit Commission.

Audit Commission (2006b) *The Planning System: Matching Expectations and Reality*. London: Audit Commission.

Baden, T. (2008) Planning to what purpose? *Town and Country Planning*, September: 369–71.

Baker, L. (2000) Lawyers forecast a log jam of appeals, *Planning*, 10 November: 1.

Baker, M., Coaffee, J. and Sherriff, G. (2007) Achieving successful participation in the new UK spatial planning system, *Planning Practice and Research*, 22(1): 79–93.

Ball, M. (1998) Institutions in British property research: a review, *Urban Studies*, 35: 1501–17.

Ball, M., Allmendinger, P. and Hughes, C. (2009) Housing supply and planning delay in the south of England, *Journal of European Real Estate Research*, 2(2): 151–69.

Banham, R., Barker, P., Hall, P. and Price, C. (1969) Non-plan: an experiment in freedom, *New Society*, 20 March.

Barker, K. (2003) *Barker Review of Housing Supply: Interim Report Analysis*. London: HM Treasury.

Barker, K. (2004) *Barker Review of Housing Supply: Final Report*. London: HM Treasury.

Barker, K. (2006a) *Barker Review of Land Use Planning: Final Report*. London: HM Treasury.

Barker, K. (2006b) *Barker Review of Land Use Planning: Interim Report*. London: HM Treasury.

Barnes, M., Bauld, L., Benzeval, M., Judge, K., MacKenzie, M. and Sullivan, H. (2005) *Health Action Zones: Partnerships for Health Equality*. London: Routledge.

Barrett, S. and Fudge, C. (eds) (1981) *Policy and Action*. London: Methuen.

Beck, U. (2002) The cosmopolitan society and its enemies, *Theory, Culture and Society*, 19: 17–44.

Better Regulation Task Force (2000) *Tackling the Impact of Increasing Regulation: A Case Study of Hotels and Restaurants*. London: Cabinet Office.

Bevir, M. (2000) New Labour: a study in ideology, *British Journal of Politics and International Relations*, 2(3): 277–301.
Bevir, M. (2005) *New Labour: A Critique*. London: Routledge.
Bevir, M. and Rhodes, R. A. W. (2006) *Governance Stories*. London: Routledge.
Booth, P. (1996) *Controlling Development: Certainty and Discretion in Europe, the USA and Hong Kong*. London: UCL Press.
Booth, P. (2003) *Planning by Consent: The Origins and Nature of British Development Control*. London: Routledge.
Booth, P. (2007) The control of discretion: planning and the common law tradition, *Planning Theory*, 6(2): 127–45.
BPF (British Property Federation) (2007) *Planning Gain Supplement Consultation*. London: BPF.
BPF (British Property Federation) (2010) *Planning Manifesto: Making Planning Work*. London: BPF.
Bramley, G., Leishman, C., Kofi Karley, N., Morgan, J. and Watkins, D. (2007) *Transforming Places: Housing Investment and Neighbourhood Market Change*. York: Joseph Rowntree Foundation.
Brenner, N. (2004) *New State Spaces: Urban Governance and the Rescaling of Statehood*. Oxford: Oxford University Press.
Brenner, N. and Theodore, N. (2002) From the 'new localism' to the spaces of neoliberalism, in N. Brenner and N. Theodore (eds), *Spaces of Neoliberalism: Urban Restructuring in North America and Western Europe*. Oxford: Blackwell, pp. 2–32.
Brindley, T., Rydin, Y. and Stoker, G. (1996) *Remaking Planning: The Politics of Urban Change in the Thatcher Years*. London: Unwin Hyman.
British Chambers of Commerce (2009) *Planning for Recovery*. London: British Chambers of Commerce.
Brown, N. and Lees, L. (2009) Young people and the regeneration of King's Cross, in R. Imrie, L. Lees and M. Raco (eds), *Regenerating London: Governance, Sustainability and Community in a Global City*. London: Routledge.
Brownill, S. (1990) *Developing London Docklands: Another Great Planning Disaster?* London: Paul Chapman.
Brownill, S. (2007) New Labour's evolving regeneration policy: the transition from Single Regeneration Budget to the single pot in Oxford, *Local Economy*, 22(3): 261–78.
Brownill, S. and Carpenter, J. (2007) Increasing participation in planning: emergent experiences of the reformed planning system in England, *Planning Practice and Research*, 22(4): 619–34.
Brownill, S. and Carpenter, J. (2009) Governance and 'integrated' planning: the case of sustainable communities in the Thames Gateway, England, *Urban Studies*, 46: 251.
Bruntland, G. (1987) *Our Common Future: Report of the World Commission on Environment and Development*. Oxford: Oxford University Press.
Bruton, M. and Nicholson, D. (1987) *Local Planning in Practice*. London: Hutchinson.
Bunnell, G. (1995) Planning gain in theory and practice: negotiation of agreements in Cambridgeshire, *Progress in Planning*, 44(1): 1–113.
Burnham, P. (2001) New Labour and the politics of depoliticisation, *British Journal of Politics and International Relations*, 3(2): 127–49.
Callinicos, A. (2001) *Against the Third Way*. Cambridge: Polity.
Cambridgeshire Horizons (2009) *Investment Package Update: The Variable Tariff/Community Infrastructure Levy and the Integrated Development Programme*. Available at: www.cambridgeshirehorizons.co.uk.

Carmona, M. (2003a) Planning indicators in England: a top-down evolutionary tale, *Built Environment*, 29(4): 281–95.
Carmona, M. (2003b) An international perspective on measuring quality in planning, *Built Environment*, 29(4): 346–66.
Catney, P., Henneberry, J. and Dixon, T. (2006) Navigating the brownfield maze: making sense of brownfield regeneration policy and governance. Paper presented at the SUBRIM conference, Imperial College, London, March.
CBI (Confederation of British Industry) (2001) *Planning for Productivity: A Ten Point Action Plan*. London: CBI.
CBI (Confederation of British Industry) (2005) *Planning Reform: Delivering for Business?* London: CBI.
CBI (Confederation of British Industry) (2007) *CBI Response on Planning-Gain Supplement, 28 February*. London: CBI.
CCC (Cambridge County Council) 2002 *Planning for Success: Cambridgeshire and Peterborough Joint Structure Plan Review*, deposit draft. Cambridge: Cambridge County Council.
CEC (Commission of the European Communities) (1990) *Green Paper on the Urban Environment*. Brussels: CEC.
Centre for Cities (2010) *Arrested Development: Are We Building Houses in the Right Places?* London: Centre for Cities.
Clifford, B. (2007) *Planning at the Coalface: The Planner's Perspective Survey – Preliminary Results*. London: King's College, Department of Geography.
Clifford, B. (2009) Planning at the coalface: British local authority planners and the experience of planning and public sector reform. Unpublished PhD thesis, King's College, London.
Cloke, P. (ed.) (1992) *Policy and Change in Thatcher's Britain*. London: Pergamon Press.
Coates, D. (2005) *Prolonged Labour: The Slow Birth of New Labour Britain*. Basingstoke: Palgrave.
Cochrane, A. (2007) *Understanding Urban Policy: A Critical Approach*. Oxford: Blackwell.
Colomb, C. (2007) Unpacking New Labour's 'urban renaissance' agenda: towards a socially sustainable reurbanization of British cities? *Planning Practice and Research*, 22(1): 1–24.
Conservative Party (2009) *Control Shift: Returning Power to Local Communities*, Policy Green Paper no. 9. London: Conservative Party.
Conservative Party (2010) *Open Source Planning*. London: Conservative Party.
Coop, S. and Thomas, H. (2007) Planning doctrine as an element in planning history: the case of Cardiff, *Planning Perspectives*, 22: 167–93.
Counsell, D. and Haughton, G (2003) Regional planning tensions: planning for economic growth and sustainable development in two contrasting English regions, *Environment and Planning C*, 21: 225–39.
Cowell, R. and Murdoch, J. (1999) Land use and limits to (regional) governance: some lessons from planning for housing and minerals in England, *International Journal of Urban and Regional Research*, 23(4): 654–69.
CPRE (Campaign to Protect Rural England) (2006) *Policy-Based Evidence Making: The Policy Exchange's War against Planning*. London: CPRE.
Crook, A. and Rowley, S. (2006) North fears southern siphon, *Planning*, 28 July.
Crook, A. and Rowley, S. (2007) Levy raises complexity fear, *Planning*, 19 January.
Cuff, N. and Smith, W. (2009) *Our Towns, our Cities: The Next Steps for Planning Reform*. London: Bow Group.
Cullingworth, J. B. (1980) Land values, compensation and betterment, *Environmental Planning 1939–1969*, Vol. IV. London: HMSO.

Davies, J. S. (2001) *Partnerships and Regimes: The Politics of Urban Regeneration in the UK*. Aldershot: Ashgate.

Davies, J. S. (2004) Conjuncture or disjuncture? An institutionalist analysis of local regeneration partnerships in the UK, *International Journal of Urban and Regional Research*, 28(3): 570–85.

Davoudi, S. and Strange, I. (eds) (2009) *Conceptions of Space and Place in Strategic Spatial Planning*. London: Routledge.

DCLG (Department of Communities and Local Government) (2005) *Planning Policy Statement 1: Delivering Sustainable Development*. London: HMSO.

DCLG (Department of Communities and Local Government) (2006a) *Planning Obligations: Best Practice*. London: HMSO.

DCLG (Department of Communities and Local Government) (2006b) *Valuing Planning Obligations: Final Report*. London: HMSO.

DCLG (Department of Communities and Local Government) (2006c) *Planning Gain Supplement: Summary of Consultation Responses*. London: HMSO.

DCLG (Department of Communities and Local Government) (2006d) *Changes to Planning Obligations: A Planning Gain Supplement Consultation*. London: HMSO.

DCLG (Department of Communities and Local Government) (2006e) *Evaluation of the Planning Delivery Grant 2005–06*. London: HMSO.

DCLG (Department of Communities and Local Government) (2006f) *Planning and Climate Change: A Supplement to PPS1 Consultation*. London: HMSO.

DCLG (Department of Communities and Local Government) (2006g) *Strategic Framework for the Thames Gateway*. London: HMSO.

DCLG (Department of Communities and Local Government) (2006h) *The Local Government White Paper: Strong and Prosperous Communities*. London: HMSO.

DCLG (Department of Communities and Local Government) (2006i) *Housing and Planning Delivery Grant Consultation Paper*. London: HMSO.

DCLG (Department of Communities and Local Government) (2006j) *Planning Policy Statement 3: Housing*. London: HMSO.

DCLG (Department of Communities and Local Government) (2007a) *Planning for a Sustainable Future: The Planning White Paper*. London: HMSO.

DCLG (Department of Communities and Local Government) (2007b) *Homes for the Future: More Affordable, More Sustainable*. London: HMSO.

DCLG (Department of Communities and Local Government) (2007c) *Housing and Planning Delivery Grant: Consultation on Allocation Mechanism*. London: HMSO.

DCLG (Department of Communities and Local Government) (2007d) *Housing and Planning Delivery Grant: Summary of Responses*. London: HMSO.

DCLG (Department of Communities and Local Government) (2007e) *Planning Fees in England: Proposals for Change*. London: HMSO.

DCLG (Department of Communities and Local Government) (2007f) *Improving the Appeals Process in the Planning System*. London: HMSO.

DCLG (Department of Communities and Local Government) (2007g) *Streamlining Local Development Frameworks*. London: HMSO.

DCLG (Department of Communities and Local Government) (2007h) *Consultation Paper on a new Planning Policy Statement 4: Planning for Sustainable Economic Development*. London: HMSO.

DCLG (Department of Communities and Local Government) (2008a) *National Indicators for Local Authorities and Local Authority Partnerships: Handbook of Definitions*, Appendix 4. London: HMSO.

DCLG (Department of Communities and Local Government) (2008b) *Spatial Plans in Practice: Supporting the Reform of Local Planning*. London: HMSO.
DCLG (Department of Communities and Local Government) (2008c) *Local Spatial Planning*. London: HMSO.
DCLG (Department of Communities and Local Government) (2009a) *Transforming Places, Changing Lives: Taking Forward the Regeneration Framework*. London: DCLG.
DCLG (Department of Communities and Local Government) (2009b) Housing: live tables. Available at: www.communities.gov.uk/housing/housingresearch/housingstatistics/live tables.
DCLG (Department of Communities and Local Government) (2009c) *Housing and Planning Delivery Grant: Consultation on Allocation Mechanism for Year 2 and 3*. London: HMSO.
DCLG (Department of Communities and Local Government) (2009d) *Planning Policy Statement 4: Planning for Sustainable Economic Growth*. London: HMSO.
DCLG (Department of Communities and Local Government) (2009e) *Community Infrastructure Levy: Detailed Proposals and Draft Regulations for the Introduction of the Community Infrastructure Levy: Consultation*. London: HMSO.
DCLG (Department of Communities and Local Government) (2009f) *Infrastructure Planning Commission: Implementation Route Map*. London: HMSO.
DCLG (Department of Communities and Local Government) (2009g) *Improving Permitted Development Consultation*. London: HMSO.
DCLG (Department of Communities and Local Government) (2010a) *Community Infrastructure Levy: Final Impact Assessment*. London: HMSO.
DCLG (Department of Communities and Local Government) (2010b) *Development Management: Proactive Planning from Pre-Application to Delivery*. London: HMSO.
DCLG (Department of Communities and Local Government) (2010c) *Improving Engagement by Statutory and Non-Statutory Consultees*. London: HMSO.
DCLG (Department of Communities and Local Government) (2010d) *Guidance on Information Requirements and Validation*. London: HMSO.
DCLG (Department of Communities and Local Government) (2010e) *Consultation on a Planning Policy Statement: Planning for a Low Carbon Future in a Changing Climate*. London: HMSO.
DEFRA (Department for Environment, Food and Rural Affairs) (2005) *Securing the Future: UK Government Sustainable Development Strategy*. London: HMSO.
DETR (Department of the Environment, Transport and the Regions) (1998) *Modernising Planning: A White Paper*. London: DETR.
DETR (Department of the Environment, Transport and the Regions) (1999) *Towards an Urban Renaissance: Final Report of the Urban Task Force*. London: HMSO.
DETR (Department of the Environment, Transport and the Regions) (2000a) *Planning Policy Guidance Note 3: Housing*. London: HMSO.
DETR (Department of the Environment, Transport and the Regions) (2000b) *Our Towns and Cities: The Future – Delivering an Urban Renaissance*. London: HMSO.
DETR (Department of the Environment, Transport and the Regions) (2000c) *Planning for Clusters: A Research Report*. London: HMSO.
DiMaggio, P. and Powell, W. (eds) (1991) *The New Institutionalism in Organizational Analysis*. Chicago: University of Chicago Press.
Dobry, G. (1975) *Review of the Development Control System: Final Report*. London: HMSO.
DoE (Department of the Environment) (1986) *The Future of Development Plans*. London: HMSO.
DoE (Department of the Environment) (1989) *The Future of Development Plans*. London: HMSO.

DoE (Department of the Environment) (1990) *This Common Inheritance: Britain's Environmental Strategy*. London: HMSO.

DoE (Department of the Environment) (1992) *Planning Policy Guidance 4: Industrial, Commercial Development and Small Firms*. London: HMSO.

Driver, S. and Martell, L. (2006) *New Labour*. 2nd edn, Cambridge: Polity.

Drivers Jonas Deloitte (2010) DJ Community Infrastructure Levy survey results. Available at: www.djdeloitte.co.uk/uk.aspx?doc=36740&sector=21445.

DTI (Department of Trade and Industry) (1999) *Biotechnology Clusters*, report of a team led by Lord Sainsbury, minister for science. London: DTI.

DTLR (Department of Transport, Local Government and the Regions) (2001a) *Planning: Delivering a Fundamental Change*. London: HMSO.

DTLR (Department of Transport, Local Government and the Regions) (2001b) *Faster, Fairer Planning for All – Byers*, DTLR press release 537, 12 December.

DTLR (Department of Transport, Local Government and the Regions) (2001c) *Reforming Planning Obligations: A Consultation Paper*. London: HMSO.

DTLR (Department of Transport, Local Government and the Regions) (2002a) *Resourcing of Local Planning Authorities*. London: HMSO.

DTLR (Department of Transport, Local Government and the Regions) (2002b) *Sustainable Communities: Delivering through Planning*. London: HMSO.

DTLR (Department of Transport, Local Government and the Regions) (2002c) *Planning Green Paper: Summary of Consultation Responses*. London: HMSO.

Edgar, E. (1983) Bitter harvest, *New Socialist*, September/October.

Edwards, M. and MacCafferty, A. (1992) 1991: a time to reflect, *Estates Gazette*, no. 9223.

Ellis, H. (2007) Does planning have a future and who cares anyway? *Town and Country Planning*, 76(1): 18–19.

Elson, M. J. (1986) *Green Belts: Conflict Mediation in the Urban Fringe*. London: Heinemann.

English Partnerships (2006) *Response to Planning Gain Supplement*. London: English Partnerships.

Evans, R. (2008) Planning and the people problem, *Journal of Environment and Planning Law*, 13. Available at: www.jplc.org/papers/RobertEvans.pdf.

Fainstein, S. (1994) *The City Builders: Property, Politics and Planning in London and New York*. Oxford: Blackwell.

Faludi, A. and Waterhout, B. (2002) *The Making of the European Spatial Development Perspective: No Masterplan*. London: Routledge.

Finlayson, A. (2009) Planning people: the ideology and rationality of New Labour. *Planning Practice and Research*, 24(1): 11–22.

Fischer, F. and Forester, J. (eds) (1993) *The Argumentative Turn in Policy Analysis and Planning*. London: UCL Press.

Flyvbjerg, B. (1998) *Rationality and Power: Democracy in Practice*. Chicago: University of Chicago Press.

Forester, J. (1989) *Planning in the Face of Power*. London: University of California Press.

Friend, J. K., Jessop, J. N. and Yewllett, C. J. L. (1974) *Public Planning: The Intercorporate Dimension*. London: Tavistock.

Fuller, C. and Geddes, N. (2008) Urban governance under neoliberalism: New Labour and the restructuring of state space, *Antipode*, 40(2): 252–82.

Gaffikin, F. and Skerrett, K. (2006) New visions for old cities: the role of visioning in planning, *Planning Theory and Practice*, 7(2): 159–78.

Gamble, A. (1984) This lady's not for turning: Thatcherism mk III, *Marxism Today*, July.

Gamble, A. (1988) *The Free Economy and the Strong State*. London: Macmillan.

Gamble, A. (2009) *The Spectre at the Feast*. Basingstoke: Palgrave.

Gamble, A. and Wright, T. (eds) (1999) *The New Social Democracy*. Oxford: Blackwell.
Gatenby, I. and Williams, C. (1996) Interpreting planning law, in M. Tewdwr-Jones (ed.), *British Planning Policy in Transition*. London: UCL Press.
Geddes, M. (2006) Partnership and the limits to local governance in England: institutionalist analysis and neoliberalism, *International Journal of Urban and Regional Research*, 30(1): 76–97.
Geddes, M. and Martin, S. (2000) The policy and politics of Best Value: currents, crosscurrents and undercurrents in the new regime, *Policy and Politics*, 28(3): 379–95.
Giddens, A. (1984) *The Constitution of Society*. Cambridge: Polity.
Giddens, A. (1998) *The Third Way*. Cambridge: Polity.
GOEE (Government Office for the East of England) (2000) *RPG6: Regional Planning Guidance for East Anglia to 2016*. London: HMSO.
Gonzalez, S. (2006) *The Northern Way: A Celebration or a Victim of the New City-Regional Government Policy?* Working Paper no. 28. London: ESRC/DCLG.
Goodstadt, V. (2009) Working across boundaries, *Town and Country Planning*, (78)3: 8.
Goodwin, M., Jones, M. and Jones, R. (2005) Devolution, constitutional change and economic development: explaining and understanding the new institutional geographies of the British state, *Regional Studies*, 39(4): 421–36.
Grant, M. (1999) Compensation and betterment, in B. Cullingworth (ed.), *British Planning*. London: Athlone Press, pp. 62–75.
Greenhalgh, P. and Shaw, K. (2003) Regional Development Agencies and physical regeneration in England: can RDAs deliver the urban renaissance? *Planning Practice and Research*, 18(2–3): 161–78.
Gunder, M. and Hillier, J. (2009) *Planning in Ten Words or Less: A Lacanian Entanglement with Spatial Planning*. Farnham: Ashgate.
Guy, S. and Henneberry, J. (2000) Understanding urban development processes: integrating the economic and the social in property research, *Urban Studies*, 13: 2399–416.
Guy, S. and Henneberry, J. (2002) Bridging the divide? Complementary perspectives on property, *Urban Studies*, 8: 1471–8.
Hajer, M. (1989) *City Politics: Hegemonic Projects and Discourse*. Aldershot: Gower.
Hajer, M. A. (1995) *The Politics of Environmental Discourse*. Oxford: Oxford University Press.
Hall, P. (2009) Planning London: a conversation with Peter Hall, in R. Imrie, L. Lees and M. Raco (eds), *Regenerating London: Governance, Sustainability and Community in a Global City*. London: Routledge.
Hall, P. (2010) Perspectives from the cycle of British planning policy, *Town and Country Planning*, 79(6): 264–7.
Hall, S. (2003) The 'Third Way' revisited: New Labour, spatial policy and the national strategy for neighbourhood renewal, *Planning Practice and Research*, 18(4): 265–77.
Harvey, D. (2005) *A Brief History of Neo-Liberalism*. Oxford: Oxford University Press.
Hastings, A. (2003) Strategic, multi-level neighbourhood regeneration: an outward looking approach at last? in R. Imrie and M. Raco (eds), *Urban Renaissance? New Labour, Community and Urban Policy*. Bristol: Policy Press.
Haughton, G. and Allmendinger, P. (2008) The soft spaces of local economic development, *Local Economy*, 23(2): 138–48.
Haughton, G. and Allmendinger, P. (forthcoming, 2010) New planning spaces post devolution, *Environment and Planning C*.
Haughton, G., Allmendinger, P., Counsell, D. and Vigar, G. (2010) *The New Spatial Planning*. London: Routledge.
Hay, C. (1999) *The Political Economy of New Labour: Labouring under False Pretences?* Manchester: Manchester University Press.

HBF (Home Builders Federation) (2010) *Let's Start at Home: Building out of Recession*. London: HBF.
Healey, P. (1992) An institutional model of the development process, *Journal of Property Research*, 9: 33–44.
Healey, P. (1994) Urban policy and property development: the institutional relations of real estate in an old industrial region, *Environment and Planning A*, 26: 177–98.
Healey, P. (1996) *Collaborative Planning: Shaping Places in Fragmented Societies*. Basingstoke: Macmillan.
Healey, P. (1998) Collaborative planning in a stakeholder society, *Town Planning Review*, 69(1): 1–22.
Healey, P. (2004a) Creativity and urban governance, *Policy Studies*, 25(2): 87–102.
Healey, P. (2004b) Towards a 'social democratic' policy agenda for cities, in C. Johnstone and M. Whitehead (eds), *New Horizons in British Urban Policy: Perspectives on New Labour's Urban Renaissance*. Aldershot: Ashgate.
Healey, P. (2006a) Transforming governance: challenges of institutional adaptation and a new politics of space, *European Planning Studies*, 14(3): 299–320.
Healey, P. (2006b) Relational complexity and the imaginative power of strategic spatial planning, *European Planning Studies*, 14(4): 525–46.
Healey, P. (2007) *Urban Complexity and Spatial Strategies: Towards a Relational Planning for our Times*. London: Routledge.
Healey, P. (2009) City regions and place development, *Regional Studies*, 43(6): 831–43.
Healey, P. (2010) *Making Better Places: The Planning Project in the Twenty-First Century*. Basingstoke: Palgrave.
Healey, P., McNamara, P., Elson, M. and Doak, A. (1988) *Land Use Planning and the Mediation of Urban Change*. Cambridge: Cambridge University Press.
Held, D. (2005) Principles of the cosmopolitan order, in G. Brook and H. Brighouse (eds), *The Political Philosophy of Cosmopolitanism*. Cambridge: Cambridge University Press.
Hillier, J. and Healey, P. (eds) (2008) *Critical Essays in Planning Theory*, Vol. 1: *Foundations of the Planning Enterprise*. Aldershot: Ashgate.
HM Government (1985) *Lifting the Burden*, Cmnd 9571. London: HMSO.
HM Government (1991) *Open Government*. London: HMSO.
HM Government (2008) *Communities in Control: Real People, Real* Power, Cm 7427. London: HMSO.
HM Government (2010) *Building the Big Society*. London: Cabinet Office.
HM Treasury (2003) *UK Membership of the Single Currency: An Assessment of the Five Economic Tests*, Cm 5776. London: HMSO.
HM Treasury (2005a) *The Government's Response to Kate Barker's Review of Housing Supply*. London: HMSO.
HM Treasury (2005b) *Planning Gain Supplement: A Consultation*. London: HMSO.
HM Treasury (2006a) Letter accompanying launch of consultation documents on PGS. London: HM Treasury.
HM Treasury (2006b) *Valuing Planning Gain: A Planning Gain Supplement Consultation*. London: HMSO.
HM Treasury (2006c) *Paying PGS: A Planning Gain Supplement Technical Consultation*. London: HMSO.
HM Treasury (2007a) *Review of Sub-National Economic Development and Regeneration*. London: HMSO.
HM Treasury (2007b) *Pre-Budget Report*. London: HMSO.
HoC (House of Commons) (2002) *Third Party Rights of Appeal*, paper 2/38, 22 May. London: House of Commons.

HoC CLGC (House of Commons Communities and Local Government Committee) (2006) *Planning Gain Supplement: Fifth Report of Session 2005–06*, Vol. 1. London: HMSO.
HoC CLGC (House of Commons Communities and Local Government Committee) (2008) *Planning Matters: Labour Shortages and Skills Gap, Eleventh Report of Session 2007–2008*. London: HMSO.
HoC HC (House of Commons Health Committee) (2004) *Third Report: Obesity*. London: HMSO.
HoC ODPM (House of Commons ODPM Housing, Planning, Local Government and the Regions Committee) (2003a) *Planning for Sustainable Housing and Communities: Sustainable Communities in the South East, 8th Report of Session*. London: HMSO.
HoC ODPM (House of Commons ODPM Housing, Planning, Local Government and the Regions Committee) (2003b) *The Effectiveness of Government Regeneration Initiatives, 7th Report of Session*. London: HMSO.
HoC PAC (House of Commons Public Accounts Committee) (2007) *The Thames Gateway: Laying the Foundations, 62nd Report of Session 2006–2007*. London: HMSO.
HoC PAC (House of Commons Public Accounts Committee) (2009) *Planning for Homes: Speeding up Planning Applications for Major Housing Developments in England, 33rd Report of Session 2008–2009*. London: HMSO.
HoC TLGR (House of Commons Transport, Local Government and the Regions Committee) (2002) *Thirteenth Report: Planning Green Paper*. London: HMSO.
Holgersen, S. and Haarstad, H. (2009) Class, community and communicative planning: urban redevelopment at King's Cross, London, *Antipode*, 41(2): 348–70.
Home Office (2009) *Working Together to Protect Crowded Places*. London: HMSO.
Hood, C. (2007) Public service management by numbers: why does it vary? Where has it come from? What are the gaps and puzzles? *Public Money and Management*, 27(2): 95–102.
Hoskins, G. and Tallon, A. (2004) Promoting the 'urban idyll': policies for city centre living, in C. Johnstone and M. Whitehead (eds), *New Horizons in British Urban Policy: Perspectives on New Labour's Urban Renaissance*. Aldershot: Ashgate.
Imrie, R. (2009) 'An exemplar for a sustainable world city': progressive urban change and the redevelopment of King's Cross, in R. Imrie, L. Lees and M. Raco (eds), *Regenerating London: Governance, Sustainability and Community in a Global City*. London: Routledge.
Imrie, R. and Raco, M. (eds) (2003) *Urban Renaissance? New Labour, Community and Urban Policy*. Bristol: Policy Press.
Imrie, R. and Thomas, H. (1999) *British Urban Policy: An Evaluation of the Urban Development Corporations*. 2nd edn, London: Sage.
Imrie, R., Lees, L. and Raco, M. (eds) (2009) *Regenerating London: Governance, Sustainability and Community in a Global City*. London: Routledge.
Inch, A. (2009) Re-evaluating street-level regulation of contradictions in planning reforms, *Planning Practice and Research*, 24(1): 83–101.
Jessop, B. (2002a) Liberalism, neoliberalism and urban governance: a state-theoretic perspective, in N. Brenner and N. Theodore (eds), *Spaces of Neoliberalism: Urban Restructuring in North America and Western Europe*. Oxford: Blackwell.
Jessop, B. (2002b) *The Future of the Capitalist State*. Cambridge: Polity.
John, P. (1999) Ideas and interests: agendas and implementation, *British Journal of Politics and International Relations*, 1(1): 39–62.
Johnstone, C. (2004) Crime, disorder and urban renaissance, in C. Johnstone and M. Whitehead (eds), *New Horizons in British Urban Policy: Perspectives on Britain's Urban Renaissance*. Aldershot: Ashgate, pp. 75–93.

Johnstone, C. and Whitehead, M. (eds) (2004) *New Horizons in British Urban Policy: Perspectives on Britain's Urban Renaissance*. Aldershot: Ashgate.

Jones, M. (2001) The rise of the regional state in economic governance: 'partnerships for prosperity' or new scales of state power? *Environment and Planning A*, 33(7): 1185–211.

Jones, M. and Ward, K. (2004) Neo-liberalism, crisis and the city: the political economy of New Labour's urban policy, in C. Johnstone and M. Whitehead (eds), *New Horizons in British Urban Policy: Perspectives on New Labour's Urban Renaissance*. Aldershot: Ashgate.

Kavanagh, D. (1994) A Major agenda? in D. Kavanagh and A. Seldon (eds), *The Major Effect*. London: Macmillan.

Kearns, A. (2003) Social capital, regeneration and urban policy, in R. Imrie and M. Raco (eds), *Urban Renaissance? New Labour, Community and Urban Policy*. Bristol: Policy Press.

Kerr, P. and Marsh, D. (1999) Explaining Thatcherism: towards a multidimensional approach, in D. Marsh, J. Buller, C. Hay, J. Johnston, P. Kerr, S. McAnulla and M. Watson, *Postwar British Politics in Perspective*. Cambridge: Polity.

Killian, J. and Pretty, D. (2007) *Planning Applications: A Faster and More Responsive System – A Call for Solutions*. London: DCLG.

Killian, J. and Pretty, D. (2008) *Planning Applications: A Faster and More Responsive System: Final Report*. London: DCLG.

King, D. S. (1987) *The New Right: Politics, Markets and Citizenship*. London: Macmillan.

King's Cross Railway Lands Group (2006) Flyer. Available at: www.kxrlg.org.uk/news/index.htm.

Knieling, J. and Othengrafen, F. (eds) (2009a) *Planning Cultures in Europe: Decoding Cultural Phenomena in Urban and Regional Planning*. Farnham: Ashgate.

Knieling, J. and Othengrafen, F. (2009b) En route to a theoretical model for comparative research on planning cultures, in J. Knieling and F. Othengrafen (eds), *Planning Cultures in Europe: Decoding Cultural Phenomena in Urban and Regional Planning*. Farnham: Ashgate.

Knight Frank (2006) *Planning Gain Supplement Audit: Final Report*. London: Knight Frank.

Kunzmann, K. R. (2009) Planning and New Labour: a view from abroad, *Planning Practice and Research*, 24(1): 139–44.

Labour Party (1996) *Planning for Prosperity*. London: Labour Party.

Lavers, A. and Webster, B. (1994) Participation in the plan-making process: financial interests and professional representation, *Journal of Property Research*, 11: 131–44.

Lawless, P. (2004) Locating and explaining area-based urban initiatives: New Deal for Communities in England, *Environment and Planning C: Government and Policy*, 22: 383–99.

Le Grand, J. (1998) The Third Way begins with Cora, *New Statesman*, 6 March, pp. 26–7.

Lee, N. (2010) Avoiding the chaos, *Town and Country Planning*, 79(6): 283–4.

Lees, L. (2003) Visions of urban renaissance: the Urban Task Force Report and the Urban White Paper, in R. Imrie and M. Raco (eds), *Urban Renaissance? New Labour, Community and Urban Policy*. Bristol: Policy Press, pp. 61–82.

Lees, L., Slater, T. and Wyly, E. (2008) *Gentrification*. London: Routledge.

Lipsky, M. (1980) *Street-Level Bureaucracy: Dilemmas of the Individual in Public Services*. New York: Russell Sage Foundation.

Local Government Association (2000) *Reforming Local Planning: Planning for Communities*. London: Local Government Association.

Lock, D. (2010) They think it's all over, *Town and Country Planning*, 79(6): 280–82.

Lord, A. (2009) The Community Infrastructure Levy: an information economics approach to understanding infrastructure provision under England's reformed spatial planning system, *Planning Theory and Practice*, 10(3): 333–49.

Lowndes, V. (1996) Varieties of New Institutionalism: a critical appraisal, *Public Administration*, 74: 181–97.

Lowndes, V. (1999) Management change in local governance, in G. Stoker (ed.), *The New Management of British Local Governance*. Basingstoke: Macmillan.

Lowndes, V. (2005) Something old, something new, something borrowed …: how institutions change (and stay the same) in local governance, *Policy Studies* 26(3): 291–309.

Lowndes, V. and Wilson, D. (2003) Balancing revisability and robustness? A New Institutionalist perspective on local government modernization, *Public Administration*, 81(2): 275–98.

Lyons, M. (2006) *National Prosperity, Local Choice and Civic Engagement: A New Partnership between Central and Local Government for the 21st Century*. London: HMSO.

McAuslan, P. (1981) *The Ideologies of Planning Law*. Oxford: Pergamon Press.

McCarthy, P. and Harrison, T. (1995) *Attitudes to Town and Country Planning*. London: HMSO.

McConnell, R. S. (1987) The implementation and the future of development plans, *Land Development Studies*, 4: 79–107.

McCormick, J. (1991) *British Politics and the Environment*. London: Earthscan.

McGregor, A. (2004) Sustainable development and warm fuzzy feelings: discourse and nature within Australian environmental imaginaries, *Geoforum*, 35: 593–606.

MacGregor, B. and Ross, A. (1995) Master or servant? The changing role of the development plan in the British planning system, *Town Planning Review*, 66(1): 41–59.

McKay, D. H. and Cox, A. W. (1979) *The Politics of Urban Change*. London: Croom Helm.

McKinsey Global Institute (1998) *Driving Productivity and Growth in the UK Economy*. London: McKinsey Global Institute.

McLean, I., Haubrich, D. and Gutierrez-Romero, R. (2007) The perils of performance measurement: the CPA regime for local authorities in England, *Public Money and Management*, 27(2): 111–17.

MacLeod, G. (2002) From urban entrepreneurialism to a 'Revanchist city'? On the spatial injustices of Glasgow's renaissance, *Antipode*, 34(3): 602–24.

March, J. and Olsen, J. (1984) *Rediscovering Institutions: The Organizational Basis of Politics*. New York: Free Press.

Marsh, D. and Rhodes, R. A. W. (1992) *Implementing Thatcherite Policies: Audit of an Era*. Buckingham: Open University Press.

Marshall, T. (2007) It works, so let's change it …, *Town and Country Planning*, September: 296–8.

Massey, D. (2005) *For Space*. London: Sage.

Massey, D. (2007) *World City*. Cambridge: Polity.

Minton, A. (2009) *Ground Control: Fear and Happiness in the Twenty-First Century City*. London: Penguin.

Moore, V. (2005) *A Practical Approach to Planning Law*. Oxford: Oxford University Press.

Morphet, J. (2008) *Modern Local Governance*. London: Sage.

Mouffe, C. (2005) *On the Political*. London: Routledge.

Murdoch, J. and Abram, S. (2002) *Rationalities of Planning*. Aldershot: Ashgate.

Nadin, V. (2007) The emergence of the spatial planning approach in England, *Planning Practice and Research* 22, 1: 43–62.

NAO (National Audit Office) (2007) *The Thames Gateway: Laying the Foundations*. London: HMSO.

NAO (National Audit Office) (2008) *Planning for Homes: Speeding up Planning Applications for Major Housing Developments in England*. London: HMSO.

Needham, B. (1996) Planning doctrine and continuity and change, *Planning Theory*, 16: 68–70.

Newman, P. (2008) Strategic spatial planning: collective action and moments of opportunity, *European Planning Studies*, 16(10): 1371–83.

NHPAU (National Housing and Planning Advice Unit) (2009) *Public Attitudes to Housing 2009*. London: HMSO.

Norton, P. and Aughey, A. (1981) *Conservatives and Conservatism*. London: Temple Smith.

ODPM (Office of the Deputy Prime Minister) (2002) *Code of Best Practice on Mobile Phone Development*. London: ODPM.

ODPM (Office of the Deputy Prime Minister) (2003a) *Sustainable Communities: Building for the Future*. London: HMSO.

ODPM (Office of the Deputy Prime Minister) (2003b) *Contributing to Sustainable Communities: A New Approach to Planning Obligations*. London: HMSO.

ODPM (Office of the Deputy Prime Minister) (2004a) *Reforming Planning Obligations: The Use of Standard Charges*. London: HMSO.

ODPM (Office of the Deputy Prime Minister) (2004b) *Community Involvement in Planning: The Government's Objectives*. London: HMSO.

ODPM (Office of the Deputy Prime Minister) (2004c) *Planning Policy Guidance Note 11: Regional Spatial Strategies*. London: HMSO.

ODPM (Office of the Deputy Prime Minister) (2004d) *Planning Policy Guidance Note 12: Local Development Frameworks*. London: HMSO.

ODPM (Office of the Deputy Prime Minister) (2004e) *Evaluation of the Planning Delivery Grant 2003–04*. London: HMSO.

ODPM (Office of the Deputy Prime Minister) (2005a) *Planning Policy Statement 1: Delivering Sustainable Development*. London: HMSO.

ODPM (2005b) *Creating Sustainable Communities: Delivering the Thames Gateway*. London: HMSO.

Ostrom, E. (1999) Institutional rational choice: an assessment of the institutional analysis and development framework, in P. Sabatier (ed.), *Theories of the Policy Process*. Boulder, CO: Westview Press, pp. 35–72.

Parker, G. (2001) Planning and rights: some repercussions of the Human Rights Act 1998 for the UK, *Planning Practice and Research*, 16(1): 5–8.

Parkinson, M., Champion, T. and Coombes, M. (2005) *State of the Cities: A Progress Report to the Delivering Sustainable Communities Summit*. London: ODPM.

PAS (Planning Advisory Service) (2007) *Planning Peer Review: Guidance for Authorities*. Available at: www.pas.gov.uk/pas/aio/41548.

PAS (Planning Advisory Service) (2008) *A Benchmark for the Spatial Planning Function*. Available at: www.pas.gov.uk/pas/aio/87347.

Peck, J. and Theodore, N. (2007) Variegated capitalism, *Progress in Human Geography*, 31(6): 731–72.

Peck, J. and Tickell, A. (2002) Neoliberalizing space, *Antipode*, 34(3): 380–404.

Peel, D. and Lloyd, M. G. (2007) Neo-traditional planning: towards a new ethos for land use planning? *Land Use Policy*, 24: 396–403.

Peel, D., Lloyd, G. and Lord, A. (2009) Business improvement districts and the discourse of contractualism, *European Planning Studies*, 17(3): 401–22.

Performance and Innovation Unit (2000) *Wiring it Up: Whitehall's Management of Cross-Cutting Policies and Services*. London: HMSO.

PINS (Planning Inspectorate) (2005) *A Guide to the Process of Assessing the Soundness of Development Plan Documents*. Available at: www.planning-inspectorate.gov.uk/pins/appeals/local_dev/ldf_testing_soundness_feb10.pdf.

Planning Officers Society (2005) *Policies for Spatial Plans: A Guide to Writing the Policy Content of Local Development Documents*. London: POS.

Planning Officers Society (2010) *The Future of Strategic Spatial Planning*. London: POS.

Powell, M. (2000) New Labour and the Third Way in the British welfare state: a new and distinctive approach? *Critical Social Policy*, 20(1): 39–60.

Poynter, G. (2009) The 2012 Olympic Games and the reshaping of East London, in R. Imrie, L. Lees and M. Raco (eds), *Regenerating London: Governance, Sustainability and Community in a Global City*. London: Routledge.

Prior, A. (2005) UK planning reform: a regulationist interpretation, *Planning Theory and Practice*, 6(4): 465–84.

Raco, M. (2005) Sustainable development, rolled-out neoliberalism and sustainable communities, *Antipode*, 37(2): 324–47.

Raco, M. and Henderson, S. (2009) Flagship regeneration in a global city: the re-making of Paddington Basin, *Urban Policy and Research*, 27: 301–14.

Radstock Action Group (2009) Planning Application: Radstock Railway Land 08/02332/RES. Available at: www.radstockactiongroup.org.uk/documents.php.

Radstock Action Group (2010) Response to BaNES draft Core Strategy. Available at: www.radstockactiongroup.org.uk/index.php.

RCEP (Royal Commission on Environmental Pollution) (2002) *Environmental Planning*. London: HMSO.

Reade, E. (1987) *British Town and Country Planning*. Oxford: Oxford University Press.

Rhodes, R. A. W. (1997) *Understanding Governance: Policy Networks, Governance, Reflexivity and Accountability*. Buckingham: Open University Press.

RICS (Royal Institution of Chartered Surveyors) (2006) *Planning-Gain Supplement: Consultation Paper*. London: RICS.

Robson, B., Bradford, M. and Deas, I. (1994) *Assessing the Impact of Urban Policy*. London: HMSO.

RTPI (Royal Town Planning Institute) (2007) *Shaping and Delivering Tomorrow's Places: Effective Practice in Spatial Planning*. London: RTPI.

RTPI (Royal Town Planning Institute) (2010) *Manifesto for Planning*. London: RTPI.

Saint Consulting (2007) *Annual Barometer of Public Attitudes towards Development*. Available at: http://tscg.co.uk.

Saint Consulting (2009) *Has the Nimbyism High Water Mark been Reached?* Available at: http://tscg.co.uk.

Sanyal, B. (ed.) (2005) *Comparative Planning Cultures*. London: Routledge.

Schmidt, V. (2008) Discursive institutionalism: the explanatory power of ideas and discourse, *Annual Review of Political Science*, 11: 303–26.

Schön, D. A. and Rein, M. (1994) *Frame Reflection: Toward the Resolution of Intractable Policy Controversies*. New York: Basic Books.

SDC (Sustainable Development Commission) (2006) *Barker Review of Land Use Planning: Call for Evidence*. London: SDC.

SEERA (South East England Regional Assembly) (2006) *Financial Impact of the Proposed Planning Gain Supplement*. Available at: www.southeast-ra.gov.uk/southeastplan/plan/pgs.html.

SEU (Social Exclusion Unit) (1998) *Bringing Britain Together: A National Strategy for Neighbourhood Renewal*. London: HMSO.

SEU (Social Exclusion Unit) (2001) *A New Commitment to Neighbourhood Renewal: National Strategy Action Plan*. London: HMSO.

Shaw, D. and Lord, A. (2007) The cultural turn? Culture change and what it means for spatial planning in England, *Planning Practice and Research*, 22(1): 63–78.

Shepley, C. (2010) A state of uncertainty, *Town and Country Planning*, 79(6): 274–7.
Simmons, M. (2008) A draft for economic disappointment? *Town and Country Planning*, March: 116–18.
Simmons, M. (2009) Inter-regional planning for the wider London region, *Town and Country Planning*, May: 236–9.
Smith, P. and Pearson, J. (2008) An environmental limits approach to spatial planning, *Town and Country Planning*, December: 509–26.
Sorenson, E. and Torfing, J. (2004) *Making Governance Networks Democratic*. Working paper, Roskilde University, Centre for Democratic Network Governance.
Sport England (2009) *Planning Play England's Future*. London: Sport England.
Stern, N. (2007) *The Economics of Climate Change: The Stern Review*. Cambridge: Cambridge University Press.
Stoker, G. (ed.) (2000) *The New Politics of British Local Governance*. Basingstoke: Palgrave.
Stoker, G. (2004) *Transforming Local Governance: From Thatcherism to New Labour*. Basingstoke: Palgrave Macmillan.
Stoker, G. (2006) *Why Politics Matters: Making Democracy Work*. Basingstoke: Palgrave Macmillan.
Stoker, G. and Wilson, D. (2004) *British Local Government into the 21st Century*. Basingstoke: Palgrave Macmillan.
Sullivan, H. and Skelcher, C. (2002) *Working across Boundaries: Collaboration in Public Services*. Basingstoke: Palgrave Macmillan.
Swyngedouw, E. (2007) Impossible 'sustainability', in R. Krueger and D. Gibbs (eds), *The Sustainable Development Paradox: Urban Political Economy in the United States and Europe*. New York: Guilford Press.
Tallon, A. (2010) *Urban Regeneration in the UK*. London: Routledge.
Taussik, J. (1992) Development plans: implications for enhanced status, *Estates Gazette*, no. 9235, 5 September.
Taylor, N. (2009) Tensions and contradictions left and right: the predictable disappointments of planning under New Labour in historical perspective, *Planning Practice and Research*, 24(1): 57–70.
TCPA (Town and Country Planning Association) (2010a) *The Future of Planning Report*. London: TCPA.
TCPA (Town and Country Planning Association) (2010b) TCPA response to open source planning, press release, 22 February.
Tewdwr-Jones, M. (2009) Governing London, in R. Imrie, L. Lees and M. Raco (eds), *Regenerating London: Governance, Sustainability and Community in a Global City*. London: Routledge.
Thornley, A. (1991) *Urban Planning under Thatcherism: The Challenge of the Market*. London: Routledge.
Thornley, A. and West, K. (2004) Urban policy integration in London: the impact of the elected mayor, in C. Johnstone and M. Whitehead (eds), *New Horizons in British Urban Policy: Perspectives on New Labour's Urban Renaissance*. Aldershot, Ashgate.
Tickell, A. and Peck, J. A. (2006) Conceptualizing neoliberalism, thinking Thatcherism, in H. Leitner, J. Peck and E. S. Sheppard (eds), *Contesting Neoliberalism: Urban Frontiers*. New York: Guilford Press.
Tiesdell, S. and Allmendinger, P. (2001) Neighbourhood regeneration and New Labour's Third Way: compromise or synergy?, *Environment and Planning C*, 19: 903–26.
Torfing, J. (2001) Path dependent Danish welfare reforms: the contribution of the new institutionalisms to understanding evolutionary change, *Scandinavian Political Studies*, 24(4): 277–309.

UCL (University College London) and Deloitte (2007) *Shaping and Delivering Tomorrow's Places: Effective Practice in Spatial Planning*. London: RTPI.
UK Government (1994) *Sustainable Development: The UK Strategy*, Cm 2426. London: HMSO.
UK Government (1998) *A New Deal for Transport: Better for Everyone*. London: HMSO.
UK Government (1999) *A Better Quality of Life: A Strategy for Sustainable Development in the United Kingdom*, Cm 4345. London: HMSO.
Underwood, J. (1980) *Town Planners in Search of a Role*. Bristol: University of Bristol, School for Advanced Urban Studies.
Upton, R. (2006) Editorial, *Planning Theory and Practice*, 7(2): 111–14.
Vaz, K. (1996) Planning for prosperity, in Home Goals: The Challenge for Labour, a report of the Talking Shop Conference of March 1996.
Vigar, G., Healey, P., Hull, A. and Davoudi, S. (2000) *Planning, Governance and Spatial Strategy in Britain: An Institutionalist Analysis*. London: Macmillan.
Walker, J. (2007) *Tariffs for Infrastructure Delivery: Building Better Communities through a Business Plan Approach*. London: Town and Country Planning Association.
Ward, S. (2004) *Planning and Urban Change*. 2nd edn, London: Sage.
Watson, J. (2009) 'Resource-hungry beast' winning few friends, *Town and Country Planning*, 78(10): 414–17.
Watson, J. and Crook, M. (2009) Fewer plans, more planning? *Town and Country Planning*, 78(3): 123–4.
While, A., Jonas, A. E. G. and Gibbs, D. (2004) Unblocking the city: growth pressures, collective provision and the search for new spaces of governance in Greater Cambridge, England, *Environment and Planning A*, 36: 279–304.
Whitehead, M. (2004) The urban neighbourhood and the new moral geographies of British urban policy, in C. Johnstone and M. Whitehead (eds), *New Horizons in British Urban Policy: Perspectives on New Labour's Urban Renaissance*. Aldershot: Ashgate, pp. 59–73.
Wilks-Heeg, S. (2009) New Labour and the reform of English local government, 1997–2007: privatizing the parts that Conservative governments could not reach? *Planning Practice and Research*, 24(1): 23–39.
Wilson, E. (2009) Multiple scales for environmental intervention: spatial planning and the environment under New Labour, *Planning Practice and Research*, 24(1): 119–38.
Zetter, J. (2009) It isn't broken but we may still have to fix it, *Town and Country Planning*, 78(6): 258–63.
Žižek, S. (2000) Holding the place, in J. Butler, E. Laclau and S. Žižek, *Contingency, Hegemony, Universality: Contemporary Dialogues on the Left*. London: Verso, pp. 308–29.

INDEX

Advisory Teams for Large Applications (ATLAS) 114, 119
ambiguity 48; of local planning objectives 85; and Planning Gain Supplement 137–8, 141, 151, 154; in policies 4, 12, 51, 63, 79, 162; in spatial planning 91, 94; of sustainable development 81; under New Labour 153; in 'urban renaissance' 17
Andrews, Baroness 15, 33, 94, 98
anti-development attitude 41, 45
anti-planning rhetoric 4, 5
antisocial behaviour 17, 68
application-based performance 119–21
Argent 70–1, 73, 75, 86–7
Audit Commission 11, 37, 115–16, 139

Barker inquiries 20, 33, 68, 163; housing supply 30, 35, 37, 52, 64, 124, 130, 133–7, 139, 145; land use planning 37, 50–1, 53,59, 85, 98, 155
Bath and North East Somerset Planning Authority (BaNES) 77–81, 84
Bellway Homes 77, 79–81
Best Value 13, 27–8, 34, 115–17, 119, 128
Better Regulation Task Force 26
Betterment Levy (1967) 132
British Property Federation (BPF) 29, 126–7, 137–9, 143, 147, 151, 165
brownfield development 25, 45, 84, 117, 123, 134–7, 144, 146, 163
brownfield targets 25, 45, 117
Bruntland Commission report 6

'burden on business' 5, 131
Business Improvement Districts 62
Business Planning Zones 27, 33, 163
Byers, Stephen 20, 27–9, 88, 94, 117

Cambridge 97–104
Cambridgeshire Horizons 98, 102–3, 148, 164
Camden Unitary Development Plan 71, 73
'cappuccino culture/policy' 52, 58
centralisation 4, 5, 13; of New Labour 13, 18, 23, 34, 50, 112, 154, 163; of Thatcherism/New Right 4–5, 8–9; of UK planning 49
children's play and planning 52
Circulars 9, 51
Citizens' Charter 1991 8, 13, 116
City Challenge 8
climate change 1, 7; and CBI 35; and housing/environment 68–70; and New Labour 2–3, 20, 23, 34–5, 51, 53; and planning 7, 12, 31, 46, 163; versus development 35
Clifton-Brown, Geoffrey 30
CO_2 emissions targets 34
coalition government 89, 108, 152–3, 165–8
'command and control' 1, 31, 34, 39, 89
Commission for Architecture and the Built Environment 64
Community Infrastructure Levy (CIL) 3, 16, 131–2, 147–51, 163
community involvement *see* involvement

Index

Community Strategies 92, 95–7, 108, 110, 158; *see also* Sustainable Community Strategies
Comprehensive Performance Assessment 115
Compulsory Competitive Tendering 8, 13, 35
Confederation of British Industry (CBI) 24, 29, 35, 53, 88, 92, 113, 119, 138, 142, 149
Conservation Areas 80, 82, 84, 116
Conservative governments 1, 4, 7, 10, 13, 23, 112, 143, 155; *see also* Major government; Thatcher governments
Conservative Party 2, 7, 16, 22, 39, 69, 152, 163–6, 168
consultation: 'front loading' of 34, 71, 87; at King's Cross 71, 73, 78; at Norton Radstock 80–2; on planning policies 124, 126, 131, 136–8, 140, 145–8; in planning process 32, 43, 97, 108; versus consensus 87
credit crunch 12, 16, 65–6, 130, 149, 155

decision-making, speed of 1, 23, 112, 117, 119
decision-taking 8, 40, 59, 98, 104, 112; *see also* development control
density targets 152, 163, 166
deregulation: as New Right policy 4–5, 8, 23; responses to 6
Development Charge (1947) 132
development control 8, 12, 22, 27–9, 31–3, 40, 59–60, 86, 90, 95, 97, 104, 110–11, 112–15, 129, 154–5, 159, 166–7; in Bath and North East Somerset (BaNES) 78, 80, 86; in Cambridge 103; and Kings Cross 76, 86; performance/speed of 115, 117, 119–27, 161, 163; powers for 102; and spatial planning 129; versus planning 36–7
Development Gains Tax (1973) 132
Development Land Tax (1976) 132
development management *see* development control
development plans 5–9, 16, 29, 31–4, 36
development planning 12, 27–9, 31–2, 40, 95, 97–8, 102–4, 113, 155, 161, 167; post-2004 system 3, 18, 29, 31, 36, 49–50, 59–60, 89, 92, 94–5, 97, 109–10, 139, 155; pre-2004 regime 95
discretion 6, 112, 154, 159, 161; in development control/UK approach 114, 121, 157; local 4–5, 8–9, 11, 13–14, 107, 112–14, 126, 163; and New

Institutionalism 58–60; of professional planners 11, 37, 44, 49, 55, 58–60; reducing/limiting/removing 5, 9, 29, 32, 59–60, 112–14, 163; and resistance 54–5
Dobry report 112
doctrine in planning 43–4, 47, 50, 55, 95

economic competitiveness 2–4, 6, 12, 34, 53–4, 64, 91, 146, 153–4, 156; and neoliberalism 65–6
economic development 20, 22, 25, 30–1, 35, 52, 59, 65–6, 81, 109, 163
Egan review of skills 37
embeddedness of planning 38, 44
English Partnerships 8, 65, 98, 102, 141, 145
Enterprise Zones (EZs) 4, 5, 9, 11, 113, 165
Environment Protection Act 1990 51
'environmental limits' 69
environmental protection 2
environmental stewardship 6, 9, 12, 153, 155
European comparisons 5, 33, 90, 93–4
European Convention on Human Rights 26
European Monetary Union 29
European Spatial Development Perspective 25, 103
experimentation 56–57, 113; by New Labour 89, 94, 163; by New Right 10–11

Falconer, Lord 15, 20, 28–30, 88, 113
fees 27, 52, 114, 118–19, 125–6
fiscal and financial incentives 4
flexibility: in planning 43–4, 58, 106–7, 112, 161; in land taxation 139, 148–50
'front loading' 34, 71, 87, 95
'fuzzy boundaries' 108, 159–60, 163
'fuzzy concepts' 41, 151

Gambling Act 2005 51
General Permitted Development orders 113
gentrification 68, 76
globalisation 31, 34–5, 37, 52, 55, 156, 157, 166
governance 1, 39–40; change in 42–60; and changing nature of state 38; 'congested' 107; corporate 23–4; issue-based 12; 'joined-up' 20, 62, 67; neoliberal spatial 156–8, 162–4, 166; networked/multi-scalar 2, 10, 14, 18–19, 22, 35, 39, 44, 52, 55, 89, 107, 160–1; networked spatial 93; New Labour's approach to 54, 67;

Index

spatial 15, 31, 33, 40, 50, 54, 155; and spatial planning 88–9; structures of 56
'governance glue' 35, 39, 162
Greater Cambridge Partnership 100
green belts 1, 18, 46, 55, 64, 99, 101
greenfield development 25, 69, 134, 137, 144, 146
'growth areas' 17, 64, 101, 104, 130, 134, 136
growth management 53–4, 87, 89–90, 102
'growth points' 130

Healey, John 152
Health Action Zones 62
Heseltine, Michael 4
HM Treasury 12, 20, 24, 26, 29–30, 35, 52, 64–5, 88–9, 91–4, 109, 131, 136, 140–1, 150
Home Builders Federation (HBF) 29, 53, 138, 143, 151, 166
the homeless 52, 62, 68
housing affordability 29, 53, 64, 68, 83, 131, 141–2, 146–7
Housing and Planning Delivery Grant (HPDG) 104, 124–5, 128–9; see also Planning Delivery Grant
Housing Information Packs (HIPs) 146
Housing Market Renewal Pathfinders 62
housing supply 30, 33, 37, 45, 52, 68, 83, 124–6, 143, 146; and Planning Gain Supplement 131–2, 134, 140, 146; in the South East 119–24
housing targets 27, 45, 59, 65, 69, 98, 104, 105, 143, 146, 154, 163
Human Rights Act 1998 26

ideology 6, 10, 18–19, 32, 54, 155, 162; of New Labour 3, 13, 17–18, 21, 23, 153–6; of Thatcherism 3–4
Improvement and Development Agency (IdeA) 114
indicators: Audit Commission 115–16; Best Value Performance Indicators (BVPIs) 13, 116–17, 119; National Indicators (NIs) 128; performance 8, 13, 40, 59–60, 114–17, 128–9, 159; planning 116
infrastructure 2, 16, 27–8, 34, 39, 59, 65, 90, 108, 130–51, 156; in Cambridge 98, 101, 104
Infrastructure Planning Commission 1, 16, 32–3, 39, 59, 65, 163, 165
institutional entrepreneurship 48–9, 55, 158–61
Integrated Spatial Strategies 95

involvement 1, 22–3, 34, 43, 47, 52, 63, 83, 85, 95
issue-based approach 12, 20, 35, 62, 158

'joined-up' governance 20, 62, 67
judicial reviews 26, 74–6, 78–9, 81, 86, 146

Kelly, Ruth 35
Killian–Pretty review 37, 85, 128, 163
King's Cross redevelopment 70–6, 85–6, 88

land taxation 130–51
land value: of brownfield locations 137; and Section 106 agreements 132, 148; uplift and taxation 130–4, 136, 139–41, 144, 151; see also Planning Gain Supplement
Licensing Act 2003 51
Local Area Agreements 62, 67, 103, 125, 129, 161, 163
Local Development Framework Core Strategies 49, 166–7
Local Development Frameworks (LDFs) 27–8, 30, 32, 45, 59–60, 69, 94–6, 108–10, 114, 128, 148, 155, 163
Local Development Orders 114, 163
local discretion 4, 5, 8, 13, 29, 60, 107, 112, 114, 126, 163
Local Economic Assessments 65
local governance 21, 42, 57, 79, 89, 92–3, 158; rule sets of 44–5
local government 9, 11, 44, 89, 91, 115–17, 135, 162; change in/reform of 18, 20–1, 31, 54, 92, 108; and New Labour 24, 44; privatisation of 8, 18, 35, 156
Local Government Act 1972 102
Local Government Act 1999 116
Local Government Act 2000 95
Local Infrastructure Funds/Groups/ Programmes 147
local plans 5, 6, 27, 165
Local Strategic Partnerships (LSPs) 52, 67, 103, 108, 158, 163
localism 7, 19, 34, 39, 152, 163–6, 168; see also New Localism
Localism Bill 152, 166
London Borough of Camden (LBC) 71, 73–5, 86
London Borough of Islington 71–4
London Borough of Merton 34
London docklands 4, 63, 105–6
Lowndes framework 44–9, 158–9

Major government 7–13, 25, 64, 86, 88, 115, 116, 160

McKinsey report 26, 37, 52, 88, 100
McNulty, Tony 30
memes 47–8, 55
'meta-governance' 39, 53
Milton Keynes 'roof tax' 132, 145
mixed-use developments 66–7, 71, 73, 77–8, 100
'modernisation': as agenda 18, 21, 36, 93–4, 115, 120; of local government 21, 31, 54; of planning 15, 22–4, 109, 159; of public sector 18
Multi-Area Agreements 62, 67, 109, 158, 161

National Audit Office reports 37, 106, 108, 127–8
National Indicators (NIs) 128
neighbourhood renewal 52, 62–3, 68
neoliberal approach 4
neoliberal spatial governance 156–8, 162–4, 166
neoliberalism 1, 4, 19, 52, 54, 60, 65–6, 91, 157–8, 167–8; definition 54; and economic competitiveness 65–6; and New Labour 17, 19, 156–8; and Third Way 41
neo-traditionalism 93–4
New Institutionalism 42–9, 57–8, 60–1, 158; versus institutionalism 44–5
New Labour (1997–2010) 13, 24–32; approach to planning 156–64; attitudes to planning 32–6; evaluations of 1–2, 17–32; and indicators/targets 114; and land taxation 148–51; pragmatism/eclecticism of 18–19; what? (influences) 17–19, 153–4; when? (periods) 20–1, 155–6; which? (sources of change) 19–20, 154–5
New Localism 34, 70, 92, 97
New Right 3–13, 19; centralising tendency 5, 8, 154; and deregulation 31, 60, 154, 156; and New Labour 23–5, 34; and role of state 48
'new times' 13, 24, 88, 92–3, 153
North America/US comparisons 5, 112
Northern Way 62
Norton Radstock redevelopment 76–87, 88, 104

obesity and planning 52, 155
objectives: multiple/contradictory 1, 17, 32, 36, 43, 49, 53, 60, 63, 65, 126, 128, 159; vague/confused 4, 60, 85, 150
'obligations creep' 142
Olympic Games 2012 105, 107

performance indicators 8, 13, 40, 59–60, 114–17, 128–9, 159
performance management 115–17
performance targets 8, 12, 21–2, 31, 34, 40, 45, 97, 113–14, 116, 118–19, 123–5, 127–9, 161, 163
permitted development 32, 113–14
'place-making' 64, 66–7, 70, 88, 93, 103, 161
plan-led approach 7, 23, 40, 50, 59–60, 76, 112–13, 154, 163, 167
plan-making 22, 95; and decision-taking (development control) 40, 102, 104, 112; and spatial planning 98, 108
planning: change in 42–60; characteristics of 42–4; 'commodification' of 114; for delivery of objectives 33; 'modernisation' of 15, 22–3, 33, 36, 93, 109, 159; meaning of 2; as neoliberal spatial governance 162–4, 166; as networked governance 35; and New Labour 1–2, 12–14; performance management in 115–17; and urban policy 62–87; versus development control 36–7
Planning Act 2008 16, 40, 147
Planning Advisory Service (PAS) 114, 119
Planning and Compensation Act 1991 7
Planning and Compulsory Purchase Act 2004 2, 29–30, 32, 40, 49–50, 59, 65, 76, 90, 98, 103, 109–110, 167; Section 38 13, 29, 59, 76, 110; Section 53 118–19; and spatial planning 89, 92, 94–6
Planning and Regulatory Services Online (PARSOL) 114
planning application fees 27, 52, 114, 118–19, 125–6
planning controls 5–6, 20, 88, 102
planning delay, analysis of 122–3
Planning Delivery Grant (PDG) 13, 29, 33, 59–60, 104, 111, 119–20, 124–9, 155, 159, 163; *see also* Housing and Planning Delivery Grant
planning doctrine 43–4, 47, 50, 55, 95
Planning Gain Supplement (PGS) 3, 12, 130–51, 154, 163, 166; tax or tariff 137
planning indicators 116
Planning Inspectorate 45, 95–7, 126
planning obligations 16, 27–8, 32, 51, 104, 130, 132–43, 145–51
Planning Performance Agreements (PPAs) 125
planning permission: and Business Planning Zones 27; in Cambridge 99; for King's Cross 71–2, 74, 76, 86; and land values 130–4, 137–8, 140, 142, 144–6; for

Norton Radstock 78, 80–2; and Planning Delivery Grant 119; refusals 74, 127, 142; and spatial planning 90; time to obtain 120–3, 165
Planning Policy Guidance Notes (PPGs) 8, 9, 80; specific notes 25, 51, 80–3, 96
Planning Policy Statements (PPSs) 85; specific statements 35, 51, 59, 65, 75, 81–2, 109, 148
Planning Portal 27
planning profession/professionals 9–10, 28, 32, 35, 37, 43–4, 53, 60, 88–9, 91, 109, 118
'planning renaissance' 15, 91
planning styles 45, 56–7, 95, 157
policy ambiguity *see* ambiguity
policy processes 43
post-political condition 54
'predict and provide' 11
Prescott, John 20, 29, 53, 64
'presumption in favour' 10–11, 45, 59, 65, 81, 112, 163, 165, 167
pricing for improved performance 125–6
privatisation: of local government 8, 18, 35, 156; of public space 63
professional autonomy/discretion 11, 14, 37, 44–5, 49, 55, 58–60
public involvement *see* involvement
public services 8, 14, 25, 33, 116
public space, privatisation of 63
public–private partnerships 63, 157

Radstock *see* Norton Radstock
redevelopment 25, 63; and housing improvement 64; of King's Cross 70–6; mixed-use 67; of Norton Radstock 76–85; and objectors 85; *see also* regeneration; 'urban renaissance'
regeneration: and consumerism 67; and land/property markets 68; New Labour approach 19, 62–6, 75–6, 87; New Right approach 8; in Thames Gateway 104–8; time and cost of 86; versus sustainable development 85; *see also* redevelopment
Regional Assemblies 68
Regional Development Agencies 1, 16, 25, 65–7, 77, 81, 163
Regional Economic Strategies 83, 95
Regional Planning Guidance (RPG) 83, 100–1
Regional Spatial Strategies (RSSs) 16, 27–8, 30, 44, 65, 69, 83, 86, 94–6, 103, 109–10, 138, 152, 155, 163, 165–6
'renaissance of planning' view 1

resistance 6, 10–11, 13, 32, 34, 45, 54, 69, 111, 124, 132, 159
resource planning 117–19
rhetorical attacks 4
Rogers report 25, 37, 63, 66
'rolling back the state' 4, 56, 164
'roof tax' 132, 139, 145
Royal Commission on Environmental Pollution 28, 34, 95
Royal Society of Chartered Surveyors (RCIS) 137–8, 146
Royal Town Planning Institute (RTPI) 31, 94–5, 97, 137, 144, 165
rule sets (Lowndes) 44–5, 47–8, 55, 61, 87, 158, 160–2
'rules in use'/'rules in form' (Ostrom) 47, 159

Sainsbury report 98–100
Saint Index 32, 110–11, 113
Schmidt framework 45–7, 49, 158–9, 162
Section 106 *see* Town and Country Planning Act 1990
self-assessment (planning gain) 141, 143
Simplified Planning Zones (SPZs) 5–6, 9, 113, 163, 165
Single Regeneration Budget (SRB) 8, 13, 64, 81, 93; for Norton Radstock 76–7
site-based performance 119–21
'smart growth' 41, 53
social exclusion 63, 68
social inclusion 3, 19, 34, 52, 64, 66, 75, 155, 157; and community 67–8
social justice 17, 20–1, 25, 33, 153, 157
South West Regional Development Agency 77, 81
spatial planning 1, 3, 11, 15, 22–4, 33–4, 36, 40–1, 46, 50–5, 58, 63, 88–111, 113, 153–4, 156–7, 159–64, 166; in Cambridge 97–104; characteristics of 90–1; and climate change 51; and development control 129; experiences of 96–7; and growth agenda 98–104; and growth management 89–90; interpretations of 91–4; as neo-traditionalism 93–4; and post-positivist theory 89; as rebranding 92; and sustainable communities 31; in Thames Gateway 98, 104–8; and 'urban renaissance' 52; vagueness of 90–1, 107
standard charging 139, 145–7
Statements of Community Involvement 34, 52
structure plans 5, 6, 108–9; Cambridgeshire 101, 103; and Norton Radstock 83

sub-regional planning 22, 74, 97, 109
sustainable communities 34, 42, 46, 63, 163; and Cambridgeshire Horizons 164; Egan review of skills for 37; and growth 130; interpretation of 70; plan/programme 57, 62, 64, 101, 105; and planning 31, 88, 131
Sustainable Community Strategies 44, 95, 110; *see also* Community Strategies
sustainable development 1–2, 6, 15, 20, 29, 31, 35, 41–2, 51, 53, 63, 65, 70, 79, 81–2, 84–6, 95, 100, 111, 126, 151, 158, 162–3, 165, 167; consensus and opposition 75–6; and spatial planning 91

target-driven approach/culture 67, 92, 97, 117
targets 28, 35, 39, 114–15, 121, 125–6, 129, 161; see also brownfield targets; CO_2 emissions targets; decision-making; density targets; housing targets; performance targets; thirteen-week target
tariffs 131–2, 137, 145, 147–8; and Community Infrastructure Levy 150–1
Tax Increment Financing 62
taxation: and infrastructure funding 139–40; of land/development 130–51; to mitigate impact 51; and reduction of supply 34; 'roof tax' 132, 139, 145
Terminal 5 (Heathrow) 27, 165
terrorism 19, 52
'test of soundness' 13, 32, 95, 97
Thames Gateway 98, 104–8
Thatcher governments 3, 5, 10, 16, 34, 88, 105, 115, 131
Thatcherism 3–4, 10, 13, 48, 153
Third Way 31–2, 41, 68, 91–3
thirteen-week target 121, 128
Town and Country Planning Act 1947 10
Town and Country Planning Act 1990 6, 51, 76, 130, 130; Section 54A 7–9, 13, 29; Section 106 74, 78, 130–3, 135–6, 138–40, 145–51

Town and Country Planning Association (TCPA) 89, 137, 165
transport 2, 25, 51, 82–5, 100–2, 108, 133, 161
Treasury *see* HM Treasury
'Triangle Site' *see* King's Cross

UK Strategy on Climate Change 51
Urban Development Corporations (UDCs) 4, 5, 9, 11, 57, 62, 66, 102, 107, 163
urban policy 62–87
urban regeneration 19, 37, 70, 113, 154, 156; new Labour's approach to 75–6
Urban Regeneration Companies 67, 105
'urban renaissance' 17, 19, 26, 41–2, 52, 53, 58, 63–4, 66–7, 85–7, 111, 131, 151, 163, 166; consensus and opposition 75–6; interpretation of 70
Urban Task Force 25, 64
US/North America comparisons 5, 112
Use Classes and orders 5, 11, 113

vagueness 47, 63, 158; of coalition proposals 153; of environmental policy 69; and King's Cross redevelopment 75–6; and New Labour 27, 153–4, 158–9, 166; of New Right 4; as opportunity for objection 85–6; and Planning Gain Supplement 137, 139–40; and 'spatial planning' 90–2, 96–7, 106–7, 109, 111; of 'sustainable communities' 65
Vaz, Keith 25, 33, 113

'Westminster model' 10, 16, 58
'win–win–win' concepts/outcomes 36, 42, 92, 151, 162

the young 62, 66, 68

zoning-based approaches/systems 5, 6, 7, 10, 59, 112–13, 167

eBooks – at www.eBookstore.tandf.co.uk

A library at your fingertips!

eBooks are electronic versions of printed books. You can store them on your PC/laptop or browse them online.

They have advantages for anyone needing rapid access to a wide variety of published, copyright information.

eBooks can help your research by enabling you to bookmark chapters, annotate text and use instant searches to find specific words or phrases. Several eBook files would fit on even a small laptop or PDA.

NEW: Save money by eSubscribing: cheap, online access to any eBook for as long as you need it.

Annual subscription packages

We now offer special low-cost bulk subscriptions to packages of eBooks in certain subject areas. These are available to libraries or to individuals.

For more information please contact webmaster.ebooks@tandf.co.uk

We're continually developing the eBook concept, so keep up to date by visiting the website.

www.eBookstore.tandf.co.uk